Stiquito™

Stiquito™

Advanced Experiments with a Simple and Inexpensive Robot

James M. Conrad
Jonathan W. Mills

IEEE
COMPUTER
SOCIETY

Los Alamitos, California

Washington • Brussels • Tokyo

Library of Congress Cataloging-in-Publication Data

Conrad, James M.
    Stiquito: advanced experiments with a simple and inexpensive robot /
James M. Conrad, Jonathan W. Mills.
        p.   cm.
    Includes bibliographical references and index.
    ISBN 0-8186-7408-3
    1. Robots—Design and construction. I. Mills, Jonathan W.
(Jonathan Wayne)   II. Title.
TJ211.C635   1998
629.8 ' 92—dc21

                                                                96-29883
                                                                CIP

IEEE Computer Society Press Order Number BP07408
Library of Congress Number 96-29883
ISBN 0-8186-7408-3

Additional copies may be ordered from:

IEEE Computer Society Press	IEEE Service Center	IEEE Computer Society
Customer Service Center	445 Hoes Lane	Watanabe Building
10662 Los Vaqueros Circle	P.O. Box 1331	1-4-2 Minami-Aoyama
P.O. Box 3014	Piscataway, NJ 08855-1331	Minato-ku, Tokyo 107-0062
Los Alamitos, CA 90720-1314	Tel: +1-908-981-0060	JAPAN
Tel: +1-714-821-8380	Fax: +1-908-981-9667	Tel: +81-3-3408-3118
Fax: +1-714-821-4641	mis.custserv@computer.org	Fax: +81-3-3408-3553
Email: cs.books@computer.org		tokyo.ofc@computer.org

Publisher: Matt Loeb
Manager of Production, CS Press: Deborah Plummer
Advertising/Promotions: Tom Fink
Production Editor: Denise Hurst
Printed in the United States of America by Victor Graphics
99 6

Contents

Preface

In 1992, Jonathan Mills announced the availability of Stiquito, a "small, inexpensive, hexapod robot." For $10 one could order from Indiana University a kit to build this small robot. Jonathan did not envision the number of requests he would receive, which by 1996 had reached more than 3,000. The volume of orders strained his personal ability to fulfill them, and he started searching for alternate suppliers for the robot, but had little success. In 1996, Jonathan was still receiving orders for Stiquito, even though he had announced that Indiana University was no longer offering the kit.

In May 1993 Jonathan also announced a call for papers. He envisioned that research papers on Stiquito applications would be published in a book. Numerous excellent papers were submitted, many of which are included here. Like all good plans, days turned into weeks, weeks turned into months, and other activities took precedence over the book. In August 1994, I joined Jonathan in the effort to prepare and organize the papers, and to identify additional material that would be valuable in the book. Discussions with publishers took longer than usual because of the unique nature of the book.

Our goal was to ensure that the book would be published *and* include a Stiquito kit. This goal made for a unique package. No other book had attempted to offer a robot, assembly instructions, and implementation examples bundled together, all for less than $50. With the support of some of the people who saw and used Stiquito, notably Jon T. Butler, a professor at the Naval Postgraduate School and an editor of the IEEE Computer Society Press, the vision of Stiquito as an educational tool received strong support.

Taking the advice that hundreds of Stiquito users have given us, our suggestion to you is simple: Read Chapter 1 and Chapter 2 fully before you attempt to build the kit. Chapter 1 contains valuable information on the objectives of the book, the contents of the kit, and intended use of the kit materials. If you read Chapter 2 completely before you build Stiquito, you will save time by avoiding common mistakes that everybody—even Jonathan and I—makes when trying to assemble the robot too hastily. You will also avoid wasting materials in your kit because of errors that others can help you prevent.

Jonathan and I would like to thank the many people who assisted in the development of this book. Jon Butler, a volunteer of the IEEE Computer Society, gave Jonathan continuous encouragement to create this book. Matt Loeb, Bill Sanders, Cheryl Smith, and Lisa O'Conner of IEEE Computer Society Press were instrumental in bringing this book to the marketplace. Of course, many thanks go to the contributing authors for their hard work and efforts for making this a success. Jonathan and I would also like to thank Klaus Peters, Dharma Agrawal, J. David Ballew, Andy Szeto, many Indiana University students, Arkansas school teachers, and all of the people who purchased Stiquito kits for providing valuable advice on the book and kit.

Jonathan wishes to thank Indiana University and its Computer Science Department for the facilities that they have provided for this project, notable in a liberal arts university that does not have an engineering school. He also wants to recognize his colleagues in the Computer Science Department, especially Steven Johnson, and his students for their support and effort that turned Stiquito from what might have been just a toy into an educational and research tool that is actively in use today.

I would like to recognize the University of Arkansas, especially my colleagues in the Computer Systems Engineering Department, as well as Collis Geren, Neil Schmitt, and Susan Vanneman, for their support of Stiquito and engineering education. Thanks also goes to Fritz Wilson and his assistants at Motorola University Support for their donations of semiconductor products. I would also like to thank my spouse, Stephanie Conrad, for her patience and support, especially when deadlines dictated that I spend more time with Stiquito than with her.

James M. Conrad

Chapter 1

An Introduction to Stiquito, the Book, and the Kit

James M. Conrad

INTRODUCTION

Welcome to the wonderful world of robotics. This book will give you a unique opportunity, one that has not been offered before, to learn about this field. It may also be the first to describe a robot *and* include the robot with the book, all for under $50. It will provide you with the skills and supplies needed to build a small robot, and will also give you instructions on how to build electronic controls for your robot.

The star of this book is *Stiquito*, a small, inexpensive hexapod (six-legged) robot. Stiquito has been used since 1992 by universities, high schools, and hobbyists. Stiquito is unique not only because it is inexpensive but also because its applications are limitless.

This introductory chapter will present an overview of robotics, the origins of Stiquito, a description of the Stiquito kit, an overview of the book, and suggestions on how to proceed with the book and building the kit.

FIRST, A WORD OF CAUTION...

This warning will be given frequently, but it is one that all potential builders must heed. Building the robot in this kit requires certain skills to produce a working robot. These hobby building skills include the following:

- Tying knots in thin metal wires
- Cutting and sanding short lengths (4 millimeters) of aluminum tubing
- Threading the wire through the tubing
- Crimping the aluminum tubing with pliers
- Stripping insulation from wire
- Patiently following instructions that require 3 to 6 hours to complete

If you think you will have difficulty with these skills or with the process of learning these skills, you should consider returning this book before opening the kit.

ROBOTICS

There are many different types of robots. The classic robots depicted in science fiction books, movies, and television shows are typically walking, talking, humanoid devices. The most useful and prevalent robot in use in the United States is the industrial arm robot used in manufacturing. You may have seen a car commercial that showed these robots welding and painting automobile bodies, for example. These robotic devices carry out repetitive and sometimes dangerous work precisely, each time. Unlike human workers, they do not need coffee breaks, health plans, or vacations (but they do need maintenance and the occasional sick day). An example of a robotic arm is the BPM Technology 3-D printer (see Figure 1.1).[1]

Figure 1.1. The BPM 3-D printer, an example of a robotic arm.

Another type of robot used in industry is the autonomous wheeled vehicle. These robots are used for surveillance or delivering goods, mail, or other supplies. They follow a signal embedded in the floor, rely on preprogrammed moves, or guide themselves using cameras and programmed floor plans. An example of an autonomous wheeled robot, which is shown in Figure 1.2, is the SR 3 Cyberguard by Cybermotion.[2] This device will travel through a warehouse or industrial building looking for signs of fire or intrusion.

Figure 1.2. The SR 3 Cyberguard by Cybermotion. A wheeled autonomous robot. Used with permission.

Although interest in walking robots is increasing, their use in industry is limited. Walking robots have advantages over wheeled robots when traversing rocky or steep terrain. One robot recently walked in the crater of a volcano and gathered data in an area too hazardous for humans to venture. Researchers at the University of Illinois have built a large walking robot, *Protobot* (see Figure 1.3), based on the physiology of a cockroach.[3] Although humor columnist Dave Barry dubbed this 2-foot creature "FrankenRoach,"[4] the robot's designers envision such devices scurrying around in hazardous environments, and even adapting to the loss of a limb.

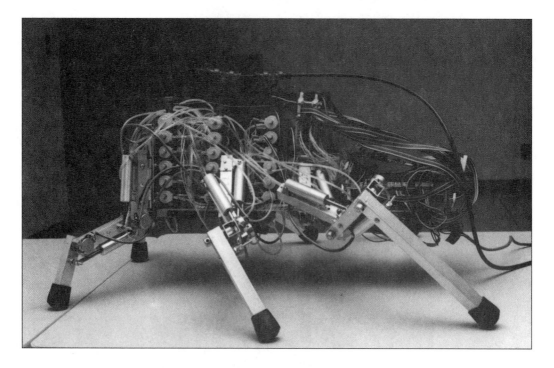

Figure 1.3. Protobot: A cockroach-inspired walking robot. From *IEEE Expert*, p. 67.

Most walking robots do not take on a true biological means of propulsion, defined as the use of contracting and relaxing muscle fiber bundles. Propulsion for most walking robots is either pneumatic or motorized. Protobot approaches a biological construction because it walks by means of pneumatic cylinders emulating antagonistic muscle pairs.

True muscle-like propulsion did not exist until recently. There is a new material, *nitinol*, that emulates the operation of a muscle. Nitinol has the properties of contracting when heated and returning to its original size when cooled; some opposable force does need to stretch the nitinol back to its original size. This new material has spawned a plethora of new, small walking robots that could not have been built with motors. Although several of these robots were designed in the early 1990s, one of them has gained international prominence because of its low cost. This robot is called *Stiquito*.

STIQUITO

In the early 1990s, Dr. Jonathan Mills was looking for a robotic platform on which to test his research on analog logic. Most platforms were prohibitively expensive, especially for a young assistant professor with limited research money. As necessity is the mother of invention, Mills set out to design his own inexpensive robot. He chose four basic materials for his designs.

1. For propulsion, he selected nitinol (specifically, Flexinol from Dynalloy, Inc.). This would provide a muscle-like reaction for his circuitry, and would closely mimic biological action. More detail on nitinol is provided throughout this book, and Appendix D contains detailed specifications for Flexinol.

2. For a counterforce to the nitinol, he selected music wire from K&S Engineering. The wire could serve as a force to stretch the nitinol back to its original length and provide support for the robot.

3. For the body of the robot, he selected ⅛-inch square plastic rod from Plastruct, Inc. The plastic is easy to cut, drill, and glue. It has relatively good heat-resistive properties.

4. For leg support, body support, and attachment of the nitinol to the plastic, he chose aluminum tubing from K&S Engineering.

Mills experimented with various designs, from a tiny, 2-inch-long, four-legged robot to a 4-inch-long robot with six floppy legs. Through this experimentation he found that the best robot movement was realized when the nitinol was parallel to the ground and the leg part touching the ground was perpendicular to the ground.

Stiquito's immediate predecessor was *Sticky*, a large hexapod robot. Sticky is 9 inches long, 5 inches wide, and 3 inches high. It contains nitinol wires inside aluminum tubes (the tubes are used primarily for support). Sticky can take 1.5-centimeter steps, and each leg has two degrees of freedom, which means that nitinol wire is used to pull the legs back as well as raise them.

Sticky was still not cost-effective, so Mills combined the concepts of earlier robots with the hexapod design of Sticky to make *Stiquito* (which means "little Sticky"). Stiquito (see Figure 1.4) has only one degree of freedom, but has a very low cost. Two years later, Mills designed a larger version of Stiquito, called *Stiquito II*, with two degrees of freedom.

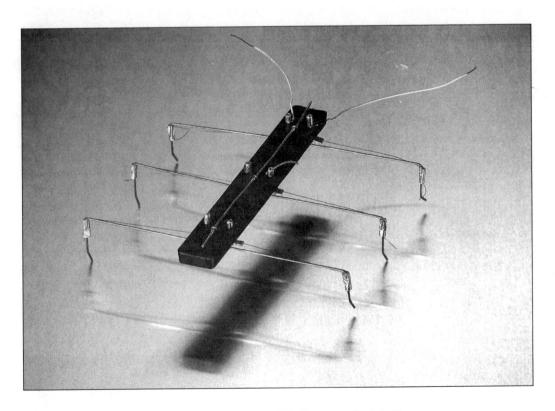

Figure 1.4. The Stiquito robot. Photograph by IEEE Computer Society Press.

At about the same time that Mills was experimenting with these legged robots, Roger Gilbertson of MondoTronics and Mark Tilden of Los Alamos Labs were also experimenting with nitinol. Gilbertson's and Tilden's robots are also described in this book.

THE STIQUITO KIT

The kit that is included with this book has enough materials to make one Stiquito robot, although there is enough extra in case you make a few errors while building the robot. The most important thing to remember when building this kit is that Stiquito is a hobby kit; it requires hobby building skills, such as cutting, sanding, and working with very small parts. For example, in one of the steps, you need to tie a knot in the nitinol wire. Nitinol is very much like thread, and it is somewhat difficult to tie a knot in it. But if you have time and patience (and after some practice), you will soon be able to tie knots like a professional.

The kit that is included with this book is a simplification of the original Stiquito described in Mills' technical report[5] that was offered as a kit from Indiana University. In your kit, the plastic Stiquito body has been premolded, so you do not have to cut, glue, and drill plastic rod to make the body. Because of this simplification, more than 10 pages were removed from the original Stiquito technical report. This new body also allows room for builders to make more errors, and requires less precision when building the robot; therefore, your robot should be more robust than earlier models.

The intent of this kit is to allow the builder to create a platform on which to start experimentation for making the robot walk. The instructions provided in Chapter 2 show how you can create a Stiquito that walks in a tripod gait; that is, it allows three legs to move at one time. What you should do is examine your goals for building the Stiquito robot and plans for controlling how Stiquito walks. If your plans include allowing each of Stiquito's six legs to be controlled independently, then you should modify the assembly of your robot so that you attach control wires to each leg individually. If the design of your robot includes putting something on top—for example, a circuit that will allow it to walk on its own—you should consider how you want it to walk. If you want it simply to walk, a tripod gait may be sufficient. If you plan to put some complex circuitry like a microcontroller on top, you may want the flexibility of being able to control all six legs.

The Stiquito robot body was also designed so it could be assembled using screws instead of aluminum crimps. If you wish to use screws instead of crimps, use the sets of holes on the body that are offset slightly. The offset holes work such that you can wrap the nitinol in the same direction as the screw thread. The nitinol is then anchored at the same distance from the legs on each side of the body. I use one brass screw ($5/16$-inch #0-80), two brass washers (#0), and two brass nuts (#0-80) for each hole. The round screw head faces down, and one of the nuts on top tightens the screw. The other nut anchors the control wire to the screw. This assembly is illustrated in Figure 1.5; see the list of suppliers in Appendix C for sets of screws, washers, and nuts. The Stiquito body is also designed so that all 12 large holes can be used for two degrees of freedom for the legs (like Sticky and Stiquito II).

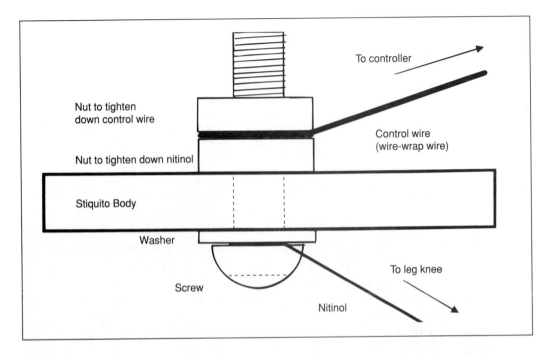

To controller

Nut to tighten
down control wire

Control wire
(wire-wrap wire)

Nut to tighten down nitinol

Stiquito Body

Washer

To leg knee

Screw

Nitinol

Figure 1.5. Using brass screws instead of aluminum crimps.

Chapter 2 has detailed assembly instructions, but here are some additional handy hints.

- This is not Lego; it is not a snap-together, easy-to-build kit. This is a hobby kit, so it takes some model-building skills. Be patient. Allow six hours to build your first robot. Jonathan swears he can build a robot in an hour, but it takes me about three (while watching sports on TV). This could be a wonderful parent-child project (in fact, my young son wants to "build bugs with daddy"). Make sure to block out enough time to complete the kit.

- Make sure you do not introduce any shorts across the control and ground wires on the robot, tether, or manual controller. Feel free to use electrical tape to insulate areas that might cause a short.

- Make sure all electrical connections are clean and free of corrosion. Sand metal parts before tying, crimping, or attaching.

- You may need to add some weight to Stiquito when using the manual controller. You can tape pennies to the bottom of the body or tape an AA battery on top. Make sure the weight does not short the control and ground wires.

In your building activities, I cannot stress enough the importance of following common safety practices.

- Wear goggles when working with the kit, as many parts of the kit can act as sharp springs.

- Use care when using a hobby knife; always cut away from you.

- Use care when using a soldering iron; watch out for burns.

- This kit is intended for adults and children over the age of 14.

BOOK ORGANIZATION

This book is organized in three sections: walking robots (chapters 2–5), control of walking robots (chapters 6–11), and research ideas (chapters 12–15). The last chapter provides a wonderful reflection on nitinol wire-based robots and their future. The following list describes the specific topics of each chapter.

- Chapter 2—Stiquito: A Small, Simple, Inexpensive Hexapod Robot. This is the starting point for building the robot kit included in the book. This chapter gives step-by-step instructions on how to assemble the kit with a manual controller.

- Chapter 3—Building Stiquito II and Tensipede. This chapter describes two cousins of the original Stiquito robot and provides step-by-step instructions on how to build them. These robots require that you purchase materials from suppliers.

- Chapter 4—Increasing Stiquito's Loading Capacity. This chapter presents another cousin of Stiquito. The SCORPIO robot is a modification of Stiquito that allows it to carry eight times more weight.

- Chapter 5—Boris: A Motorless Six-Legged Walking Machine. Boris can be considered the inspiration for Stiquito. Boris is larger and more robust than Stiquito. This chapter provides well-written, step-by-step instructions on how to build Boris.

- Chapter 6—A PC-Based Controller for Stiquito Robots. This chapter describes an interface that allows you to use an IBM PC or compatible computer to control the actuators on Stiquito II or Tensipede and experiment with various gaits. Concepts from this chapter can be used to design a simple PC controller for Stiquito.

- Chapter 7—An M68HC11 Microcontroller-Based Stiquito Controller. A design is presented for an autonomous Stiquito controlled by a Motorola M68HC11 microcontroller. The system also includes a handheld remote control to send gait instructions to the robot.

- Chapter 8—An M68HC11-Based Stiquito Colony Communication System. An M68HC11-based controller was designed with the ability to send and receive infrared signals. The new design incorporates the idea of a colony containing a queen and several colonists. The queen generates movement commands and transmits infrared movement signals to the colonists.

- Chapter 9—A General-Purpose Controller for Stiquito. This chapter describes a general-purpose controller for the Stiquito robot. The controller has the ability to control the direction of walking, can synchronize the leg movements, and can be programmed for different gaits.

- Chapter 10—SCORPIO: Hardware Design. The goal of the SCORPIO robotics project is to develop a microcontroller system for walking robots to perform independent, intelligent operations. This chapter covers the hardware aspects of the SCORPIO design.

- Chapter 11—SCORPIO: Software Design. This chapter describes the software of SCORPIO, including the kernel, a control language, an on-board interpreter, and the control/sensing routines.

- Chapter 12—Łukasiewicz' Insect: The Role of Continuous-Valued Logic in a Mobile Robot's Sensors, Control, and Locomotion. This reprinted article describes how Stiquito can be controlled via an analog computer.

- Chapter 13—Stiquito, a Platform for Artificial Intelligence. This chapter presents examples and opportunities for studying and applying the paradigms of artificial intelligence using Stiquito. Topics addressed include genetic algorithms, emergent cooperation, and neural networks.

- Chapter 14—Cooperative Behaviors of Autonomous Mobile Robots. This chapter describes a simulator designed to show Stiquito robots playing the game *Hunt the Wumpus*. The simulation tests cooperative behavior concepts of the robots.

- Chapter 15—The Simulation of a Six-Legged Autonomous Robot Guided by Vision. This chapter describes a computer graphics simulation of a six-legged autonomous robot that wanders inside a maze guided solely by what it sees through a built-in camera located in its "forehead."

- Chapter 16—The Future for Nitinol-Propelled Walking Robots. A wonderful chapter about the possible uses of nitinol wire. A must-read chapter for ideas on what to do after you build Stiquito.

- Appendix A—Author Biographies. Meet the authors of the book chapters, and find out how to contact them.

- Appendix B—An Analog Driver Circuit for Nitinol-Propelled Walking Robots. This appendix contains a schematic and brief instructions on how to build an analog controller. The controller is tailored for the Stiquito robot, but can be extended to other robots that use nitinol.

- Appendix C—Sources of Materials for Stiquito. This appendix lists some suppliers of Stiquito parts and electronics. It also lists suppliers of other robotic parts and kits.

- Appendix D—Technical Characteristics of Flexinol Actuator Wire. This appendix is the technical report from Dynalloy on its Flexinol (nitinol) wire. This will be useful for those who wish to further investigate this wire.

EDUCATIONAL USES OF STIQUITO

Although this book is available for use by any hobbyist or engineer, one of the main functions of this book is to provide an environment for educators to introduce robotics and robotic control to students. This book can be used at the high school or university level to introduce students to the concepts of analog electronics, digital electronics, computer control, and robotics. Since this book comes complete with robot and assembly instructions, it can easily serve as a required textbook for a class, with only a minimal amount of electronics required to investigate the other areas.

For example, you might want to start with the Stiquito robot and the manual controller and investigate how the robot works with respect to basic electrical fundamentals: current, voltage, and resistance. You can then expand on that concept by introducing analog electronics. You can build a simple computer controller similar to the design introduced in Chapter 6, which requires only a printed circuit board, a D-shell connector, and a couple of transistors. You can further explore electronics by designing and building a circuit similar to that described in Appendix B. Advanced classes can continue by working on digital controllers using microprocessors and microcontrollers.

The Stiquito robot also provides opportunities to examine other engineering disciplines. For example, previous exercises have included creating Stiquito kits similar to what is found in this book using the raw products. Given long lengths of music wire, long lengths of aluminum tubing, spools of wire, and bags of other parts, students can be asked to fill a small plastic bag with the parts necessary to later build Stiquito. This requires students to contemplate industrial engineering concepts, including assembly lines and materials handling. A description of this exercise can be found on the IEEE Computer Society Press Stiquito Web page.

Most of the authors in this book have used Stiquito in classes. Many of the chapters were generated by student projects guided by professors Jonathan Mills, John Estell, Susan Mengel, and James Conrad. Additional experiments were conducted using Stiquito as the basis to learn circuits. Often, when studying a topic, students are only provided the theory, written assignments, and basic experiments. They are not given an opportunity to build an actual application or implementation, something they can call their own and perhaps take with them. The Stiquito robot can be used in these cases.

As with any project conducted in a school setting, you will need additional supplies in case students break their robot kits. Contact IEEE Computer Society Press to purchase additional kits, or contact some of the suppliers listed in the back of this book for repair materials.

WHAT'S NEXT?

By now you may be asking, "So how should I proceed with this book?" Hobbyists and engineers will be prone to just jump in and hack at the kit. I strongly recommend that you do some reading first, then plan your course of action.

If you are a reflective person, the type who has to have all the facts before you proceed, by all means read the entire book. Twice. You should then have a good idea of what has been done and what works. You will be well prepared to build Stiquito and implement an existing controller, or even design your own. Your design will most likely work correctly the first time.

If you are an impatient person, the type who starts working without reading the instructions, I suggest you at least read *something.* You are an experimenter, and will figure out how to build Stiquito in time, but at least read Chapter 2. This will save you time and money in the end. But you might have more fun in the meantime just digging in.

If you are a slightly impatient, yet slightly cautious person (like most of us), your plan of action is simple. You should read a few of the chapters, choose the controller you want to build, build Stiquito and the controller, and then read the remaining chapters. This plan of attack will give you a good basis for Stiquito and some simple control, from which you can design and build other Stiquito robots and controllers. For example, after reading chapters 2 and 6, and Appendix B, you could build and test Stiquito with the manual controller. Then you could design and build a PC-based controller that is simpler than the design presented in Chapter 6 (control a Stiquito that is hardwired for a tripod gait).

Of course, you can always roll up your sleeves and start with the manual controller Stiquito. Then, you can buy another kit and build the PC-based controller and Stiquito. Then you can buy another kit and build the analog controller Stiquito. Then you can buy...well, additional kits can be purchased from the publisher.

The important point to keep in mind is that Stiquito provides an inexpensive robotic platform for experimentation. Its control circuitry and possible applications are limited only by your imagination.

Figure 1.6. A favorite comic for Stiquito labs. Used with permission.

REFERENCES

1. For more information, see the Stiquito home page at http://www.computer.org/books.

2. The home page address for Cybermotion, Inc., SR 3 Cyberguard, is http://www.cyber-motion.com/~cyberdog.

3. Price, Dick. 1995. Climbing the walls. *IEEE Expert*, April, pp. 67–70.

4. Barry, Dave. 1995. Dr. FrankenRoach, *The Washington Post Magazine*, p. 32.

5. Mills, Jonathan W. 1992. Stiquito: A small, simple, inexpensive hexapod robot. Technical report 363a, Computer Science Department, Indiana University.

Chapter 2

Stiquito: A Small, Simple, Inexpensive Hexapod Robot

Jonathan W. Mills

INTRODUCTION

Legged robots are typically large, complex, and expensive. These factors have limited their use in research and education. Few laboratories can afford to construct 100-legged robot centipedes, or 100 six-legged robots to study emergent cooperative behavior; few universities can give each student in a robotics class his or her own walking robot.

A small, simple, and inexpensive six-legged robot that addresses these needs is described in this chapter. The robot is 75 millimeters long, 70 millimeters wide, 25 millimeters high, and weighs 10 grams. It is constructed of fewer than 40 parts, 12 of which move: Six legs bend in response to six nitinol actuator wires. Nitinol wire, tradenamed Flexinol, is an alloy of nickel and titanium that contracts when heated. It is also called shape-memory alloy.[1,2] Most parts of the robot perform more than one electrical or mechanical function, but the design can be easily modified. For example, pairs of legs and actuators can be replicated to produce a mechanical centipede with flexible joints between leg segments.

The robot is intended for use as a research and educational platform to study computational sensors,[3,4] subsumption architectures,[5] neural gait control,[6] behavior of social insects,[7] and machine vision.[8] The robot can be powered and controlled through a tether, or autonomously with on-board power supply and electronics.[9] It is capable of carrying up to 50 grams while walking at a speed of 3 to 10 centimeters per minute over slightly textured surfaces such as pressboard, indoor-outdoor carpet, or poured concrete. The feet can be modified to walk on other surfaces. The robot walks when the heat-activated nitinol actuator wires attached to the legs contract. Heat is generated by passing an electric current through the nitinol wire. The legs can be actuated individually or in groups to yield tripod, caterpillar, or other gaits. The robot is named Stiquito after its larger and more complex predecessor, Sticky.

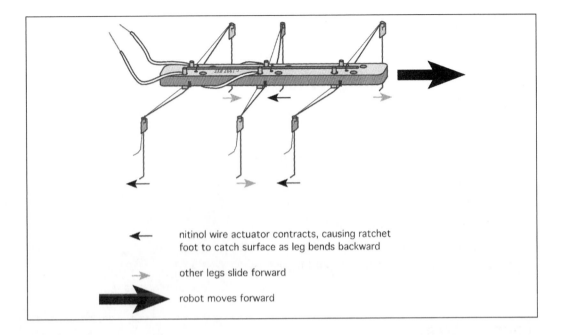

Figure 2.1. How the robot walks.

PREPARING TO BUILD STIQUITO

All materials and tools must be on hand before building the robot. Check the materials in the kit against the parts list. If anything is missing, contact the vendor. The tools needed to construct the robot are typically available in hardware stores, electronics supply stores, or hobby shops. A clear workspace and a relaxed frame of mind will be helpful during construction, especially when installing the nitinol actuators. Correct installation of the actuators will result in a robot that walks well, while a sloppy job will almost certainly lead to one that barely twitches.

Materials List

The robot should come with the following materials.

Amount	Part number	Item
1 each	ST-100	Molded plastic Stiquito body
1 each	ST-101	Molded plastic manual controller
2 x 100 mm	K&S Engineering No. 100	$^1/_{16}''$ outside diameter aluminum tubing
5 x 100 mm	K&S Engineering No. 499	0.020″ music wire
600 mm	Dynalloy 0.004″ Dia. 70 C	0.004″ (100µm) nitinol wire (Flexinol®)
70 mm	generic	20 AWG copper hook-up wire
600 mm	generic	28 AWG copper wire-wrap wire
1,500 mm	generic	34 AWG copper magnet wire
1 each	generic	9-volt terminal assembly
1 each	generic	320-grit sandpaper
1 each	generic	600-grit sandpaper

To complete the robot, the following must be purchased.

Amount	Part number	Item
1 each	generic	9-volt battery

Tools List

The following tools are needed to build this robot.

- Needlenose pliers
- Wire cutters
- Small knife (X-Acto type)
- Ruler graded in millimeters
- Voltmeter or two AA batteries and holder

PRECAUTIONS

- Always follow the manufacturer's instructions when using a tool.
- Wear safety glasses to avoid injury to eyes from broken tools, or pieces of plastic or metal that might fly away at high velocity as a result of cutting or sawing. Be especially careful when cutting music wire with wire cutters (cut wire can be forced into a finger, for example).
- Use motorized tools, such as Dremel Moto-tools,™ carefully. Motorized tools are not needed to construct Stiquito, although they are helpful if many robots will be built.
- Use a piece of pressboard, dense cardboard, or a cutting board, to protect the work surface, if necessary.

REQUIRED SKILLS

Please remember that assembling Stiquito requires hobby-kit-building skills. This section has been included for the benefit of readers who have not assembled kits before. Practice the skills needed to build Stiquito before assembling the robot, using scrap plastic and thin wire. The kit has extra material in case of mistakes, but there is not enough with which to practice.

Measuring—Following the adage "measure twice, cut once" will prevent most mistakes. Use any metric ruler graded in millimeters. Many figures in this chapter are life-size, so any dimensions included can be used to measure parts.

Cutting—Before cutting, check that your fingers are not in the way of the knife, and that a slip of the knife will not damage anything nearby. Direct the knife away from yourself to avoid injury. Make small cuts to avoid removing too much material or making too large or deep a cut.

Deburring—Sawing and cutting can leave rough edges, or burrs, on some parts. Remove the burrs by sanding the rough edge, trimming the burr with a small knife, or lightly abrading the part with a drill bit held in a pin vise. Leaving burrs on parts, especially crimps, can cause the nitinol actuators to break. Parts that are press-fitted can bend or break during assembly if not deburred.

Sanding—The ends of the aluminum tubing should be deburred by sanding them with fine (320-grit) sandpaper. Lightly sand nitinol and music wire with ultrafine (600-grit) sandpaper or emery paper to remove oxide. Sand the wire after it is bent or knotted to avoid breaking it. Sanding wire too much can weaken it enough to break during assembly or when the robot is operating.

Knotting and Crimping Nitinol Wire—Nitinol is similar to stainless steel. The 0.004-inch nitinol wire used in this robot can be knotted without breaking the wire as long as the knot is not tightened excessively. Knotting and crimping nitinol wire is the most reliable way tested to attach the actuators. Nitinol actuators must be taut, and attached so they cannot pull loose, if this robot is to walk well. The knot-and-crimp attachments have proven reliable for over 300,000 cycles (approximately 100 hours of continuous walking).

Of the other ways to anchor nitinol actuators, a U-shaped bend in the nitinol wire can pull out of a crimp far enough to reduce leg motion; soldering is difficult to control because the wire contracts and can lose its "memory"; soldering and epoxying nitinol wire might not hold under repeated actuation; and pinning or screwing the nitinol wire to the small parts used in the robot is more complex than knotting and crimping. To tie a knot, make a loop in the wire, run one end of the wire through the loop to make an overhand knot, then pull by hand to decrease the diameter of the knot's loop to about 3 millimeters. Slide the knot nearly to the end of the wire using a length of stiff wire, then grasp the end of the wire nearest the knot with the needlenose pliers, and, holding the other end of the wire (or the crimp if one is attached to the other end) with your hand, pull sharply several times with the pliers to tighten the knot. The knot is tight enough when a small loop, approximately 0.1 millimeters in diameter, remains. Tightening the knot further can break the wire.

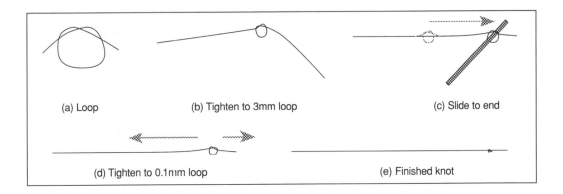

(a) Loop   (b) Tighten to 3mm loop   (c) Slide to end

(d) Tighten to 0.1mm loop   (e) Finished knot

Figure 2.2. Knotting nitinol.

Crimps are hollow connectors that are squeezed shut to hold, attach, or connect one or more objects. This robot uses short lengths of aluminum tubing as crimps to anchor the knotted nitinol actuator wires securely. Two types of crimps are needed. Body crimps hold the actuator wire alone. They are press-fitted into holes in the robot's body to attach the actuator wires indirectly to the control wires; there is no electrical connection between the actuator wire and the plastic body. Leg crimps hold both the actuator wire and the music-wire leg; they attach directly and electrically connect the actuator wire to the music-wire leg.

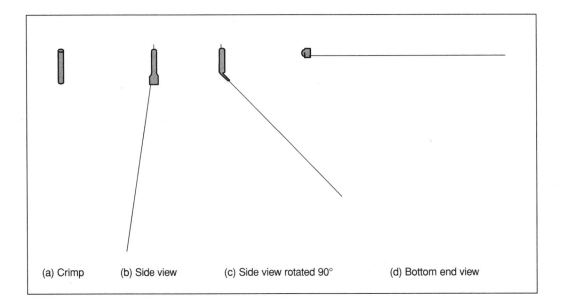

(a) Crimp   (b) Side view   (c) Side view rotated 90°   (d) Bottom end view

Figure 2.3. Body crimp (views with respect to body crimp).

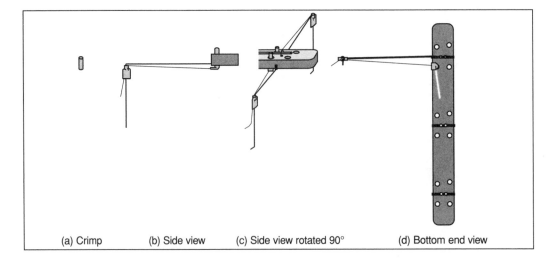

(a) Crimp   (b) Side view   (c) Side view rotated 90°   (d) Bottom end view

Figure 2.4. Leg crimp (views with respect to leg crimp).

Fixing Mistakes—No matter how carefully one works, mistakes do happen. Most are easy to fix, because almost all steps in the construction of Stiquito allow some tolerance, except for tensioning the nitinol actuator wires, where no slack is allowable. Here are some common problems, and how to work around them.

- *Music wire bent incorrectly.* Ninety-degree bends or greater can be rebent gently once or twice before breaking the wire. Bends less than 90 degrees, such as the 15-degree V-clamp in the legs, will usually break if re-bent. Stiquito will work with four legs if more music wire cannot be found. Stiquito will work with 0.015-inch-diameter music wire, but cannot carry much weight.

- *Knot in wrong place.* Tie a new knot and keep going. It might be preferable to cut the nitinol in 70-millimeter lengths instead of 60-millimeter lengths for this reason. Untying tight knots in nitinol usually breaks the wire.

- *Crimp must be removed or replaced.* Crimps can be gently squeezed across the wide dimension to undo them, but they should then be discarded. Extra tubing is provided to make new crimps. It will probably be necessary to redo the leg crimps once or twice until you get the hang of tensioning the actuator wires.

CONSTRUCTING STIQUITO

Stiquito has four major assemblies: the body, the legs and power bus, the control wires, and the actuators. The actuators are made of nitinol wire.

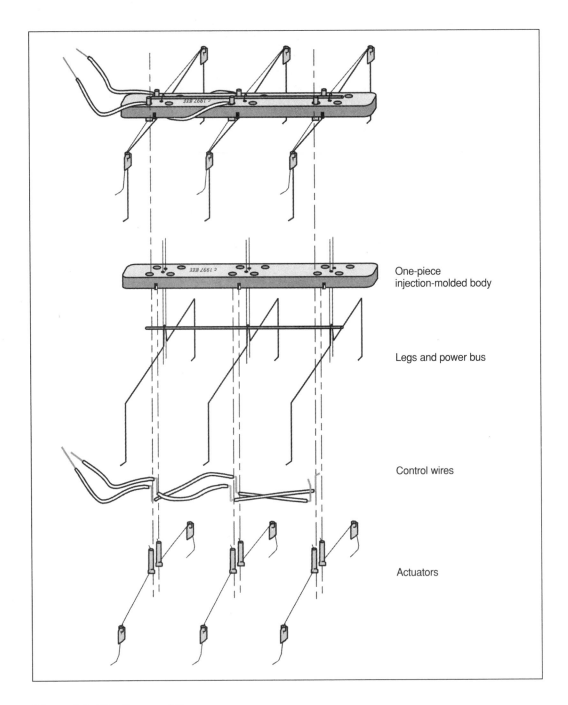

Figure 2.5. Stiquito assemblies.

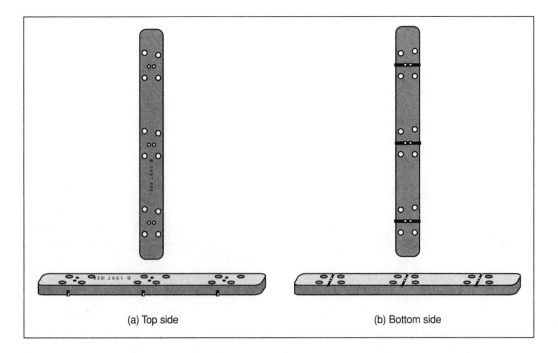

(a) Top side   (b) Bottom side

Figure 2.6. The molded body.

The Body

The body provides structural strength and locates the attachment points for the legs and the nitinol actuator wires.

The body is molded with holes and grooves. Examine the plastic to ensure that every hole goes all the way through and that there are no rough edges.

The plastic body has 18 holes. The smallest set of six holes is used to attach the legs to the body. The set of six large holes that are parallel to the small holes is used to assemble the leg actuators. The following instructions use only these 12 holes.

The other set of six holes that have three holes slightly offset can be used for other purposes, including:

- Assembling Stiquito with screws and nuts, as mentioned in Chapter 1.

- Assembling Stiquito with legs that have two degrees of freedom, similar to Stiquito II described in Chapter 3.

- Mounting a circuit board for controlling Stiquito, like the circuit described in Appendix B.

These extra holes add to the flexibility of Stiquito for experimentation.

The Legs and Power Bus

The legs are assembled in pairs from three 100-millimeter lengths of 0.020-inch music wire. The music-wire legs perform three functions.

1. *Support.* The legs support the weight of Stiquito and its battery and control electronics. Because the wire is bent to fit into a leg clip groove molded in the body, each leg in the pair is mechanically isolated.

2. *Power distribution.* All legs share a common electrical power connection to the power bus and route current to the nitinol actuator wires. The V-bend in the music wire clamps the power bus to the top of the body and electrically connects it to the legs.

3. *Recovery force.* The music wire acts as a leaf spring to provide recovery force for the nitinol wire actuator. Without this spring, or if the actuator is attached loosely, the nitinol wire will contract, but will fail to return to its original extended length.

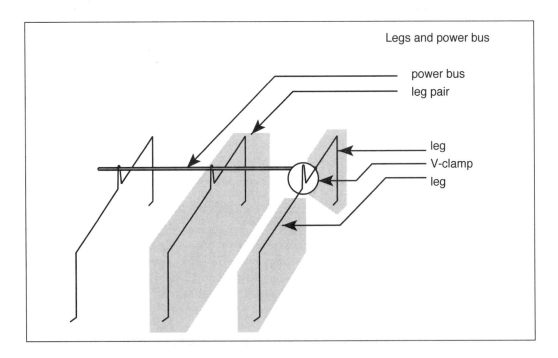

Figure 2.7. Legs and power bus.

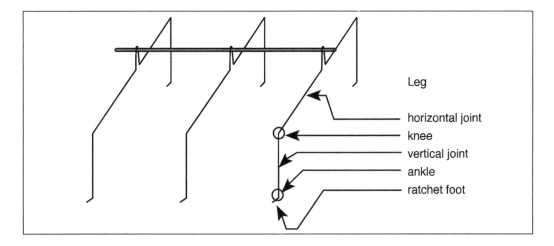

Figure 2.8. Leg detail.

Begin assembling the legs by using three 100-millimeter lengths of 0.020-millimeter music wire. Bend each music wire in the middle to a 15-degree angle. Do not bend the wires too far, or they might crack or break. The apex that forms the V-clamp should be rounded, not sharp. Lightly sand the inside of each V-clamp with the 600-grit sandpaper to remove oxide.

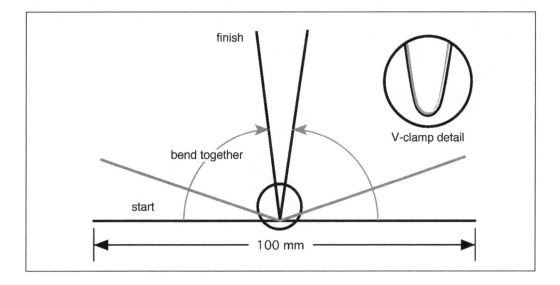

Figure 2.9. Bending the legs (enlargement 6X).

Remove the 70-millimeter length of 20 AWG copper wire from the kit. This is the power bus.

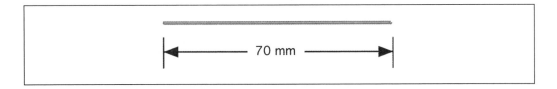

70 mm

Figure 2.10. Power bus.

Lay the power bus along the top of the body between the leg holes. Temporarily clamp the power bus by bending the legs together, inserting them through the leg holes, and then pulling the legs through from the other side until the power bus is held tightly by the V-clamp.

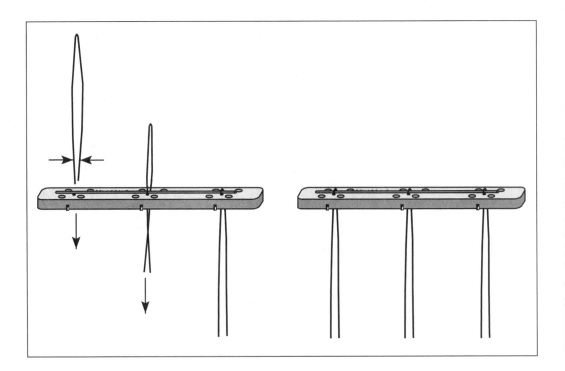

Figure 2.11. Temporarily clamping power bus with all the legs (all enlargements 2x).

Turn the body over and permanently clamp the power bus, simultaneously attaching the legs, by spreading each leg in a pair outward by hand, while at the same time pulling upward on the legs. When the legs are almost horizontal, grasp each leg in turn with the needlenose pliers and firmly bend it downward, while continuing to pull outward, until the leg lies in the clip groove.

At this point the power bus should be securely clamped in place. Check to ensure that it will not touch any of the body crimps; insert the length of aluminum tubing and check with a voltmeter. Also check that the electrical connection between the power bus and the legs is good. It should be less than 2 ohms.

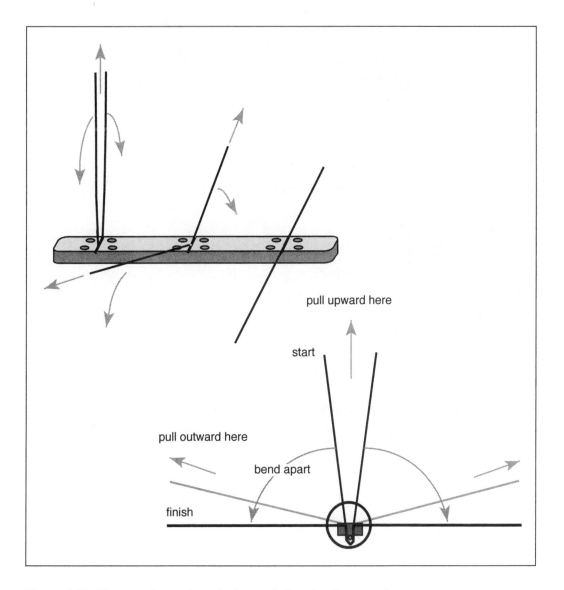

Figure 2.12. Permanently attaching the legs and clamping the power bus.

Adjust the legs so that they are in a horizontal plane with the bottom of the body and are parallel to each other.

Working with the bottom of the body facing up, form the knee, which separates the horizontal joint from the vertical joint of each leg, by bending the music wire 90 degrees about 30 millimeters from the edge of the body. Adjust the vertical joints so they are parallel.

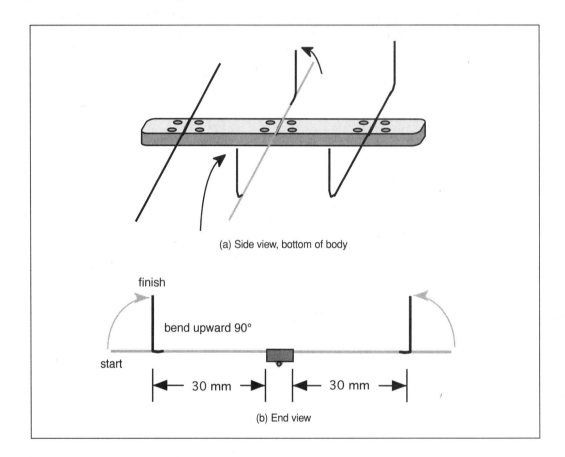

(a) Side view, bottom of body

(b) End view

Figure 2.13. Forming the knee and the horizontal and vertical joints.

Trim the vertical joints using the wire cutters so that all legs touch the ground. This completes assembly of the legs and power bus.

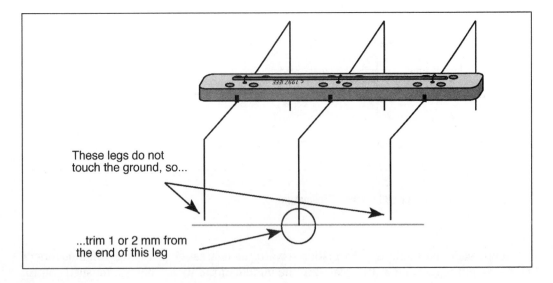

These legs do not
touch the ground, so...

...trim 1 or 2 mm from
the end of this leg

Figure 2.14. Trimming the vertical joints.

Do not bend the music wire to make the ratchet feet now. Wait until the actuators have been completed and tested.

The Control Wires

Cut two 120-millimeter lengths of 28 AWG copper wire-wrap wire from the kit for control wires. Prepare the control wires for a hardwired tripod gait by stripping 12 millimeters of insulation from each end using the hobby knife.

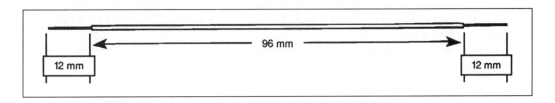

96 mm

12 mm

12 mm

Figure 2.15. Control wire.

Next, separate two lengths of insulation, leaving them on the wire. Use a knife to make a cut in the insulation 36 millimeters from one end of the wire. Make a second cut in the insulation 66 millimeters from the same end. Cut all the way through the insulation, all around the wire. Do not cut the wire.

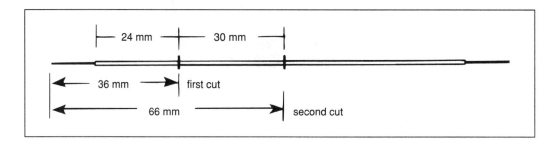

Figure 2.16. Separating two lengths of insulation.

Gently slide the sections of insulation toward the nearest end, leaving 4 millimeters of bare wire at the end and at two places in the middle of the wire. This is a finished control wire. Make two of them.

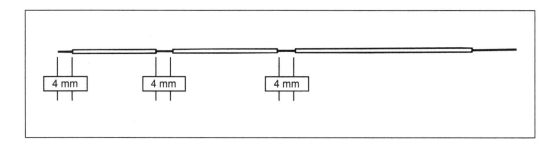

Figure 2.17. Finished control wire.

Do not install the control wires now. The control wires will be installed when the actuators are attached.

The Actuators

Stiquito is small and simple because it uses nitinol actuator wires.

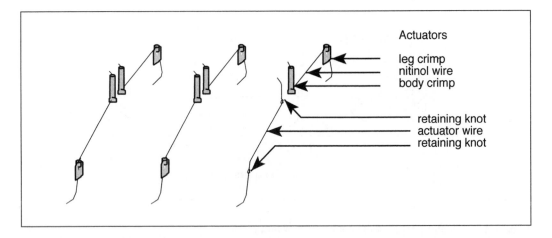

Figure 2.18. Actuators.

The nitinol wire translates the heat induced by an electric current into mechanical motion, replacing stepping motors, screws, and other components otherwise needed to make a leg move. The mechanical motion results from changes in the crystalline structure of nitinol. The crystalline structure is in a deformable state (the martensite) below the martensite transformation temperature, M_t. In this state the wire's length can change by as much as 10 percent. The nitinol wire is purchased as an expanded martensite (that is, a *trained* wire).

When the wire is heated above the austenite transformation temperature A_t (1 in Figure 20), the crystalline structure changes to a strong and undeformable state (the austenite). As long as the temperature of the wire is kept slightly above A_t, the wire will remain contracted. During normal use of the nitinol wire, a recovery force, or tension, is applied while it is an austenite.

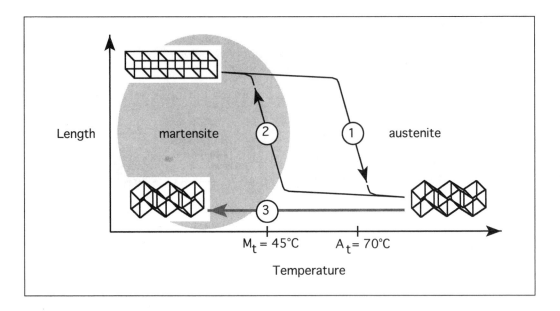

Figure 2.19. Changes in crystalline structure of nitinol.

When the temperature falls below M_t the austenite transforms back into the deformable martensite (2), and the recovery force pulls the wire back into its original, expanded form. If no recovery force is applied as the temperature falls below M_t, then the wire will remain short as it returns to the martensite (3), although it can recover its original length by cycling again while a recovery force is applied. If the wire is heated too far above A_t, then a new, shorter length results upon transformation to the martensite; the "memory" of the original, longer length cannot be restored.

Nitinol wire will operate for millions of cycles if it is not overheated and if a suitable recovery force is applied during each transformation. Stiquito's manual controller prevents overheating if used as directed. Autonomous controllers must limit the current supplied to the nitinol actuator wires to avoid overheating them. The music-wire legs provide the correct recovery force.

The actuators, legs, and power bus combine to route power, provide the recovery force, and support the robot.

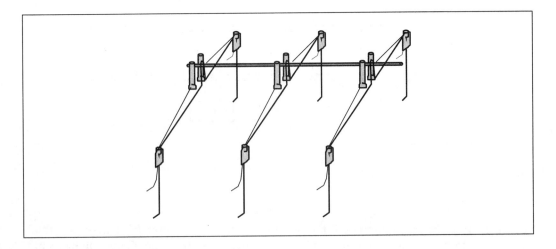

Figure 2.20. Actuators, legs, and power bus.

The following steps are needed to build the actuators, attach them to the legs and body, and form the ratchet feet.

Cut Nitinol Wires to Size—Begin making the actuators by cutting six 60-millimeter lengths of nitinol wire (there will be extra nitinol wire left over). It might be preferable to use some of the extra nitinol in the kit, cutting six 70-millimeter lengths to make it easier to tie the retaining knots during installation of the leg crimps.

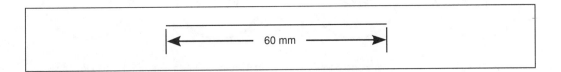

Figure 2.21. Nitinol wire.

Make Body and Leg Crimps—Next remove the 100-millimeter length of aluminum tubing from the kit. It will be used to make the leg and body crimps. Using the knife, cut six 9-millimeter body crimps and six 4-millimeter leg crimps from the aluminum tubing. Sand the ends of the crimps, then run the end of the knife through them to deburr the ends.

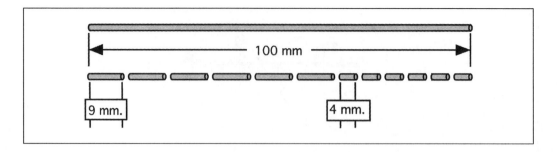

Figure 2.22. Body and leg crimps.

Attach Body Crimps to Nitinol Wires—Attach a body crimp to each length of nitinol wire. Tie a retaining knot in one end of the wire. Using the 600-grit sandpaper, lightly sand the nitinol wire at the knot to remove oxide and improve the electrical connection to the body crimp.

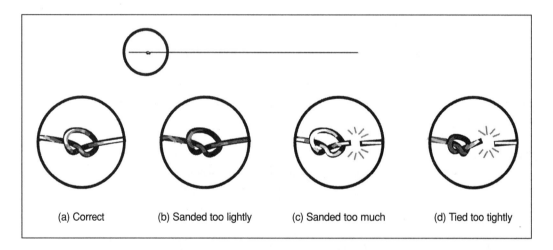

(a) Correct   (b) Sanded too lightly   (c) Sanded too much   (d) Tied too tightly

Figure 2.23. Tying and sanding retaining knot (all enlargements 10x).

Select a 9-millimeter body crimp, insert the knotted end of the wire into the body crimp, and pull it through the crimp; the knot must extend out the other end.

Figure 2.24. Knotted end of nitinol wire inserted into body crimp.

Turn the crimp until the nitinol wire is at either the right or left side of the crimp, then, using the needlenose pliers, squeeze shut about 2 millimeters of the body crimp at the end opposite the knot. The unknotted wire should protrude from either the right or left side of the flattened part of the crimp, not the middle. Pull the knotted end back into the crimp until the retaining knot catches in the crimped end.

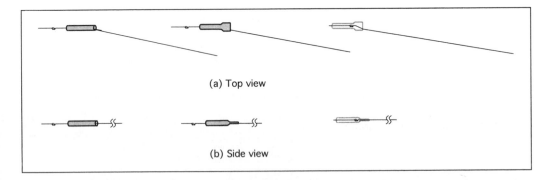

(a) Top view

(b) Side view

Figure 2.25. Attaching body crimp to nitinol wire.

Orient three body crimps so the nitinol wire exits from the left side of the flat, crimped end, and three so the nitinol wire exits from the right side. This ensures that the nitinol wire will protrude as far to the rear as possible from the music-wire leg when the actuator is attached to the body.

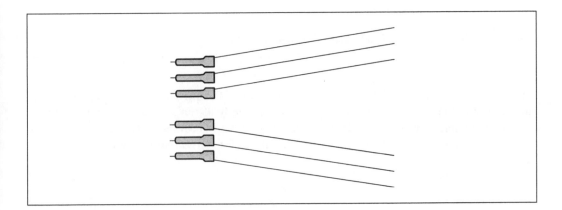

Figure 2.26. Left- and right-oriented body crimps.

Next, retaining the left- and right-handed orientation of each body crimp, grasp the flat end of each body crimp with the needlenose pliers and bend it upward to a 45-degree angle.

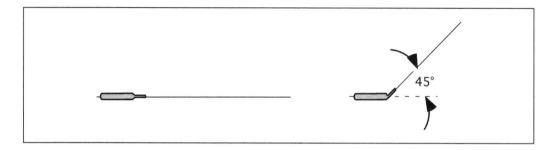

Figure 2.27. Bending the body crimp.

There should now be three left and three right body crimps with nitinol wire attached.

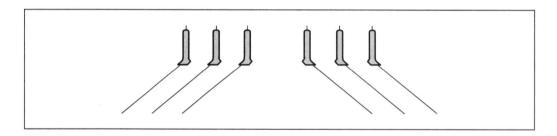

Figure 2.28. Finished left- and right-handed body crimps.

Insert Control Wires and Body Crimps into Body Crimp Holes—Viewing the body with the bottom side facing up and the front of the body at the top, use left-handed body crimps on the left side of the body and right-handed body crimps on the right side. This will put the attachment point for each actuator wire as far to the rear of the leg as possible.

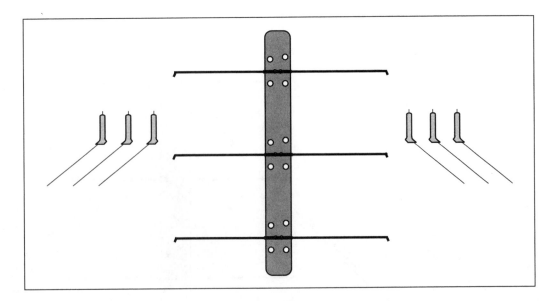

Figure 2.29. Orienting body crimps to bottom side of the body (front of body at top).

Beginning at the front of the body, insert the end of one control wire so that the bare wire enters the front-left body crimp hole at the V-groove.

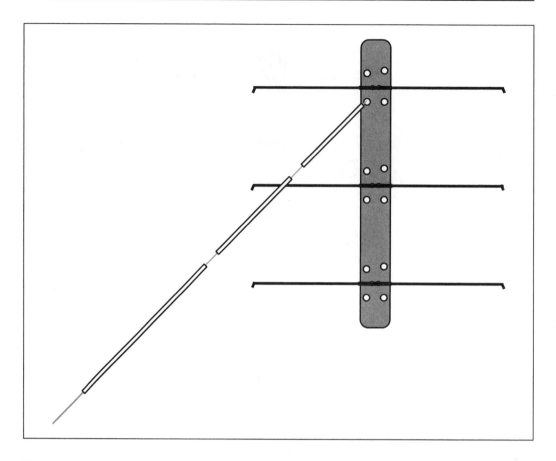

Figure 2.30. Inserting control wire.

Orient and insert a left-handed body crimp into the body crimp hole, securing the control wire.

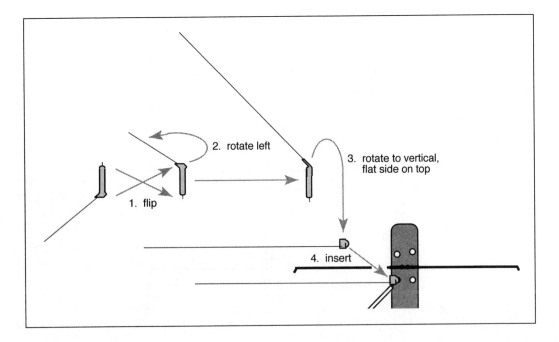

Figure 2.31. Attaching body crimp.

Continue to string the control wire and secure it with body crimps, hardwiring a tripod gait. The control wire should be run through the front-left, middle-right, and rear-left body crimp holes.

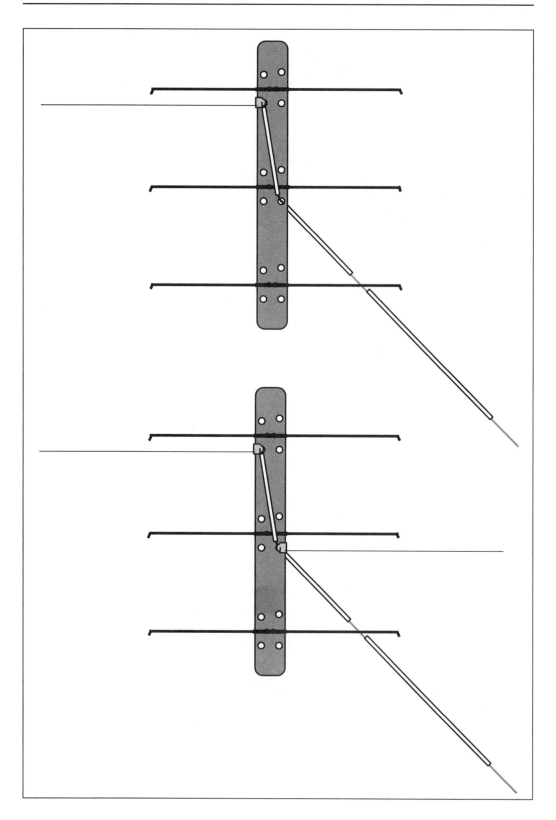

Figure 2.32. Securing control wire with right-handed body crimp in middle-right body crimp hole.

When installing the rear crimp, wrap the control wire around the body so that it enters the body crimp hole from the bottom side of the body.

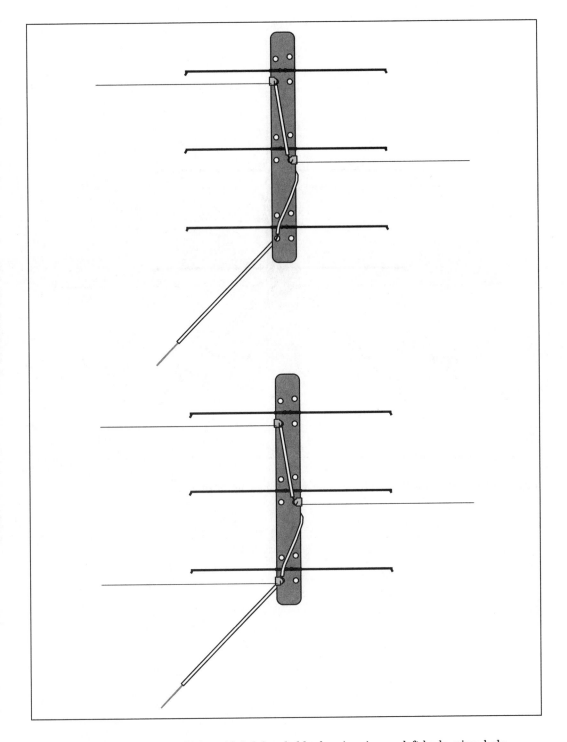

Figure 2.33. Securing control wire with left-handed body crimp in rear-left body crimp hole.

Following a similar sequence, install the second control wire in the opposite holes.

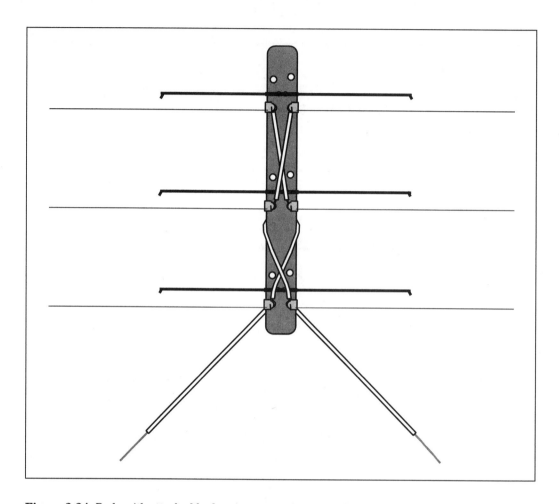

Figure 2.34. Body with attached body crimps securing control wires.

Finish the control wire and body crimp assembly by bending the 45-degree tabs on the body crimps flat against the body using the needlenose pliers. This causes the finished nitinol actuator wires to pull the leg backward and slightly downward during operation.

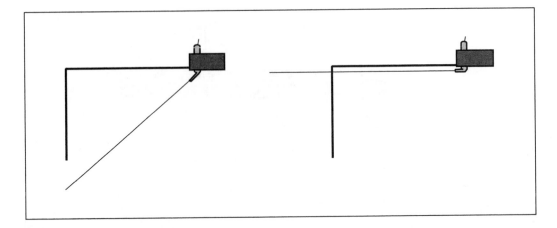

Figure 2.35. Finishing the body crimps (end view).

Attach Leg Crimps—For each nitinol wire actuator, position a knot at the knee near the vertical joint by tying an overhand knot around the vertical joint. Tighten the knot to a 0.1mm loop with the pliers.

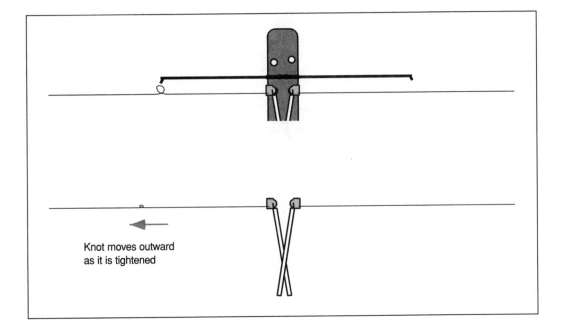

Knot moves outward
as it is tightened

Figure 2.36. Locating leg crimp knot.

Sand the knot and the vertical joint near the knee with 600-grit sandpaper. Select a 4-millimeter leg crimp and slip it onto the vertical joint and slide it up to the knee.

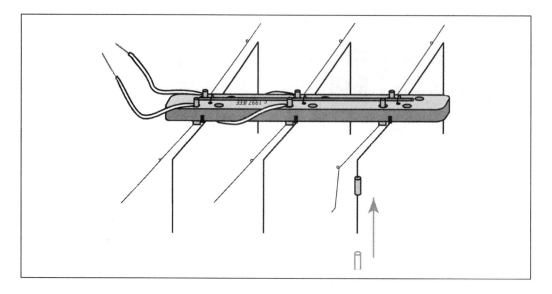

Figure 2.37. Slip crimp onto leg.

Thread the nitinol wire through the crimp.

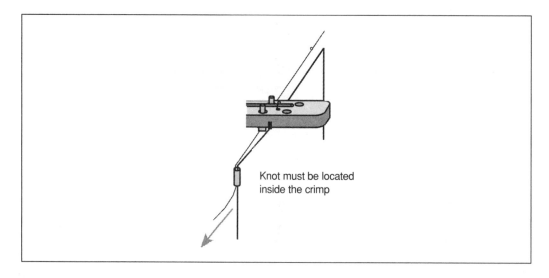

Figure 2.38. Threading wire through crimp.

Pull the wire taut so there is a slight backward bend in the horizontal joint. The slight bend ensures there is no slack in the nitinol actuator wire. There is not enough tension if there is no bend in the leg. There is too much tension if the leg is bent more than 2 millimeters backward at the knee.

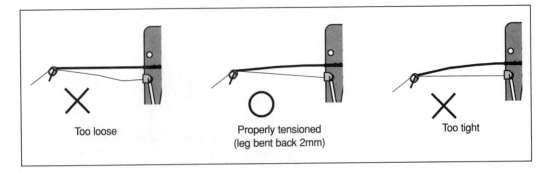

Figure 2.39. Correct tension (O) and incorrect tension (X).

The nitinol actuator wire should enter the crimp at the point nearest the body. The knot should be inside the crimp. When these conditions are met, use the needlenose pliers to make a loose crimp by squeezing the bottom 3.5 millimeters of the crimp just enough to catch the knot.

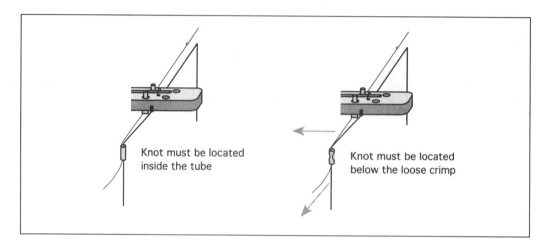

Figure 2.40. Making the loose crimp.

Crimping changes the tension in the actuator wire and might allow it to become slack. Adjust the tension while the crimp is loose by pulling on the wire (the horizontal joint should be bent slightly backward). When the actuator wire is taut, hold it in place and squeeze the crimp shut tightly.

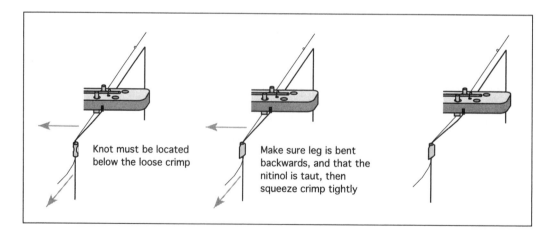

Figure 2.41. Making the tight crimp.

Measure the resistance from the leg near the clip groove to the body crimp where it protrudes above the top of the body. The initial resistance should be between 5 and 7 ohms. The resistance will increase to between 15 and 25 ohms as the connections age. Test the operation of the actuators by applying current from two 1.5-volt AA cells at the leg near the body and the body crimp for no more than half a second to prevent the actuator wire from overheating. The leg should immediately bend backward 3 to 7 millimeters as measured at the vertical joint, and then return to its original position.

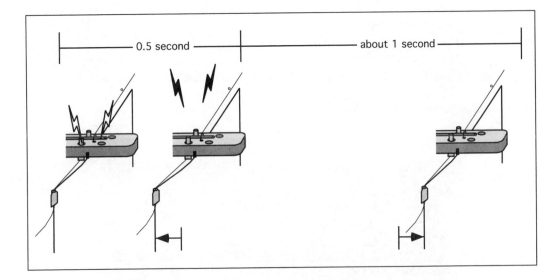

Figure 2.42. Testing the actuator.

Test the leg and actuator assembly after installing each leg crimp.

If the test is successful, continue to attach and test the remaining leg crimps.

Figure 2.43. Finished actuators (side view).

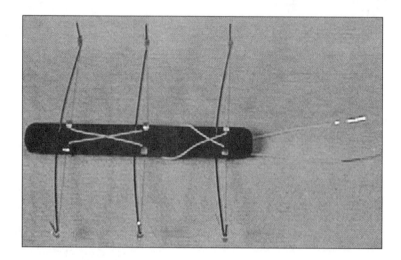

Figure 2.44. Finished actuators (bottom view).

The Ratchet Feet—Be sure to test the actuators before making the ratchet feet in case the leg crimps must be replaced. Form an ankle and ratchet foot by bending the tip of each vertical joint backward and slightly outward. The ratchet foot should be about 2 millimeters long and make a 110-degree angle downward from the vertical joint. The ankle should face toward the front of the robot.

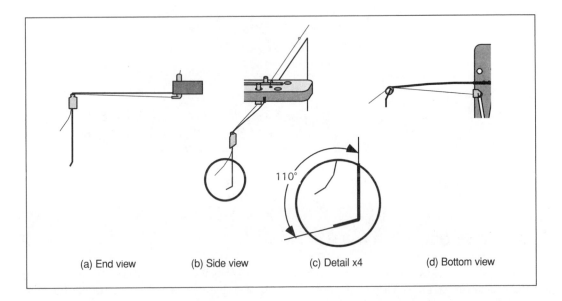

Figure 2.45. Ratchet foot.

This completes the robot. The next section describes how to manually control Stiquito.

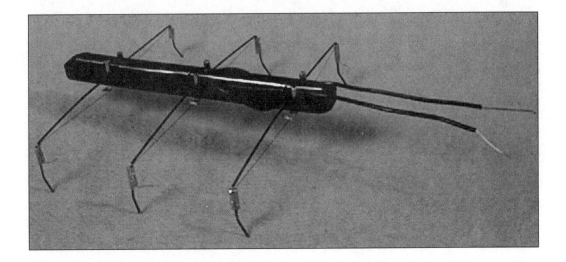

Figure 2.46. Side view of finished robot.

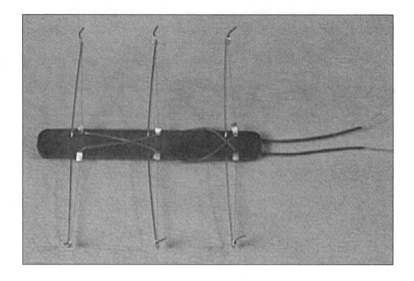

Figure 2.47. Finished actuators (bottom view).

TAKING STIQUITO FOR A WALK

Stiquito is hardwired to walk in a tripod gait. Other gaits can be hardwired by removing the body crimps and rewiring the control wires. Each leg can be wired separately for more complex gaits, especially when using a hardware controller on an autonomous robot.[9]

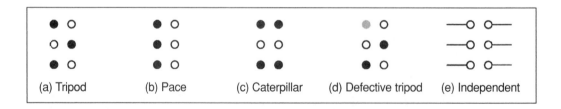

Figure 2.48. Gaits.

A simple manual controller to test Stiquito's walk is described in this section. Following that is a troubleshooting section, which contains the symptoms and causes of some common problems that might be encountered, based on experience with this robot.

Manually Operated Controller

A manual controller is a good way to test hardwired gaits. It is simple and almost foolproof. If used as directed it will prevent overheating that could damage the nitinol actuator wires. The manual controller is a pair of normally open switches with a terminal to attach a 9-volt cell. Pressing the switches alternately will power the legs, causing Stiquito to walk. If Stiquito walks with the manual controller, then the control circuits described in later chapters should work equally well.

Make the manual controller by starting with the provided plastic handpiece.

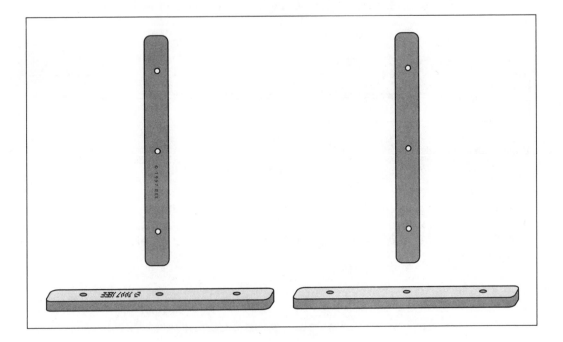

Figure 2.49. The molded plastic manual controller handpiece.

Cut two 120-millimeter lengths of 28 AWG copper wire-wrap wire from the kit. Strip 12 millimeters of insulation from each end.

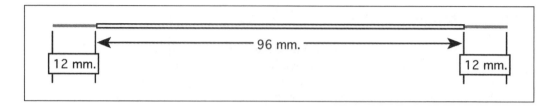

Figure 2.50. Control wire.

Bend one end of each wire in half, leaving a sharp V-bend of bare wire 6 millimeters long.

Figure 2.51. V-bend.

Remove the second 100-millimeter piece of aluminum tubing from the kit. Cut three 9-millimeter crimps. Two will be contact crimps, and one will be a power crimp. Cut six 4-millimeter connecting crimps. Approximately 40 millimeters of kit tubing remains for spare crimps.

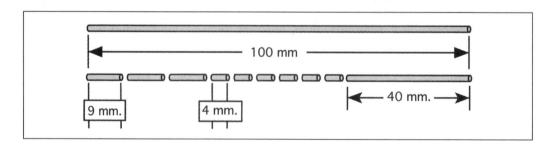

Figure 2.52. Contact, power, and connecting crimps (spare tubing shown).

Make two contact crimp assemblies. Into each of two 9-millimeter crimps insert a V-bend until the tip protrudes slightly beyond the crimp. Using the needlenose pliers, crimp 2 millimeters of the end with the slightly protruding V-bend.

Figure 2.53. Contact crimp assemblies.

Bend the flat part of the crimp to a 45-degree angle.

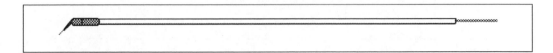

Figure 2.54. 45-degree bend in contact crimp.

Next make one power crimp assembly. Select the 9-volt terminal. If necessary, remove 5 millimeters of insulation from the end of each wire attached to the terminal. Insert one of the wires into the remaining 9-millimeter crimp all the way to the insulation and crimp 2 millimeters at the end nearest the insulation.

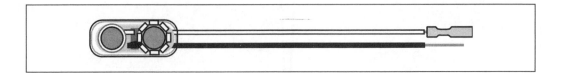

Figure 2.55. Power crimp assembly.

Take one 85-millimeter length of 0.020-inch music wire from the kit. Bend the wire in the center to a 15-degree angle to make a V-clamp.

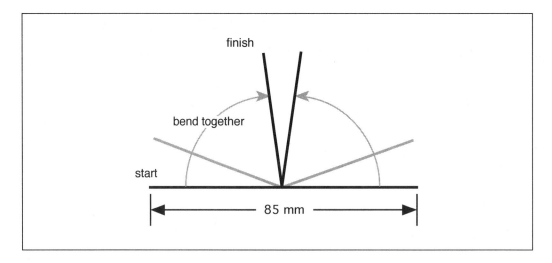

Figure 2.56. V-clamp.

Insert the music wire into the center hole in the plastic handpiece. Make sure there is some space between the wires near the bend.

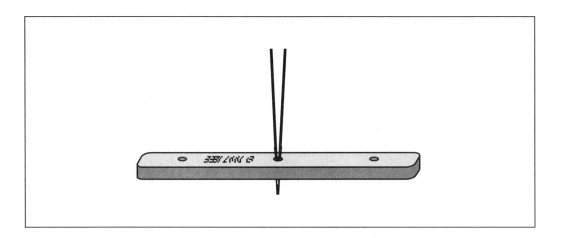

Figure 2.57. Insert music wire.

Insert the power crimp between the wires at the bend of the music wire. Center the crimp at the bend.

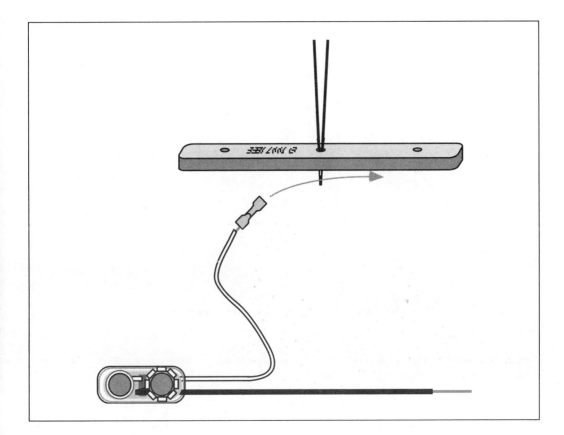

Figure 2.58. Insert power crimp.

Place the manual controller with the power crimp down on a hard surface. Firmly press down on the plastic handpiece with your thumbs, as close to the music wire as possible. Next, bend the music wire to the outside of the handpiece.

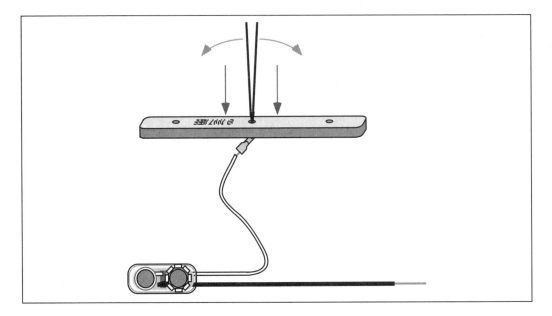

Figure 2.59. Bending the V-clamp outwards. Press down firmly to catch crimp tightly in "V" while bending music wire outward.

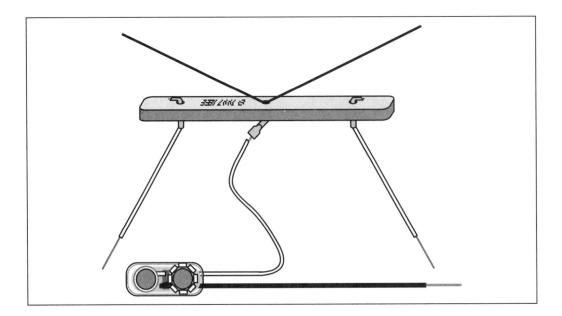

Figure 2.60. Insert and bend contact crimp assemblies.

Bend each end of the V-clamp back on itself to prevent poking yourself in the finger when using the controller.

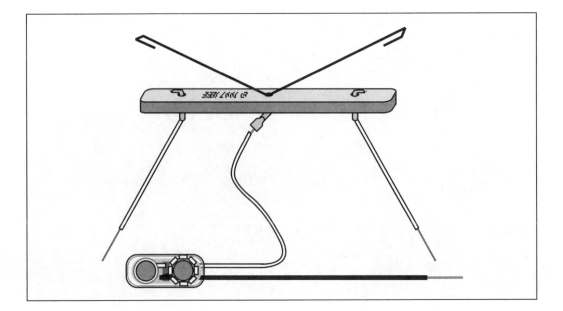

Figure 2.61. Bend the V-clamp ends.

The external arms of the V-clamp form the two manual actuators of the double switch. The bend of the V-clamp inside the handpiece connects the power crimp to the double switch.

This completes the manual controller.

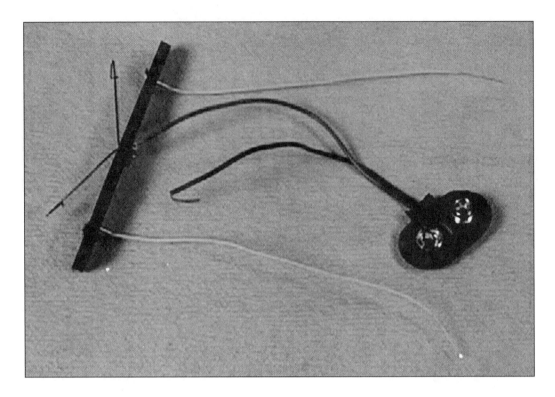

Figure 2.62. Completed manual controller.

To use the manual controller to make Stiquito walk, cut three 500-millimeter (50-centimeter) lengths of 34 AWG magnet wire. Sand the insulation from 14 millimeters of each end of the magnet wire.

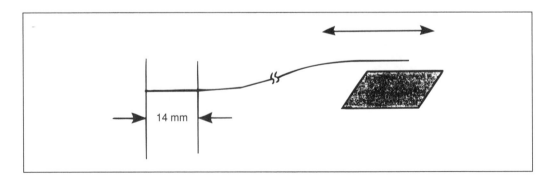

14 mm

Figure 2.63. Remove insulation from magnet wire.

Use the six connecting crimps to attach the bare magnet wire to the ends of Stiquito's power bus and control wires, and to the ends of the manual controller's wires. Attach a 9-volt cell to the terminal, then briefly and alternately press each arm of the double switch. Stiquito should walk with a tripod gait. If it does not, see the next section for troubleshooting suggestions.

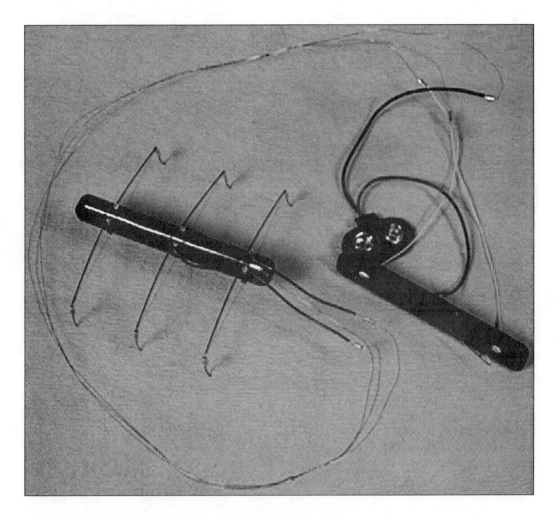

Figure 2.64. The finished Stiquito robot and manual controller.

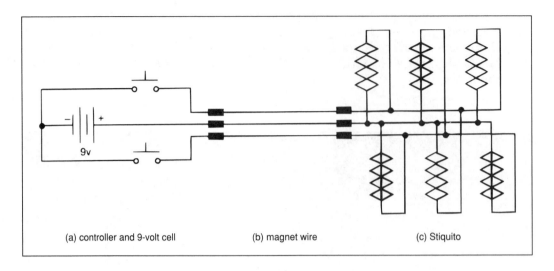

(a) controller and 9-volt cell   (b) magnet wire   (c) Stiquito

Figure 2.65. Controller, magnet wire, and Stiquito circuit schematic.

Troubleshooting

Troubleshooting is applied logical deduction. To avoid the frustration encountered when a project fails to work as expected, *expect it not to work*. Take this attitude from the start, and think about factors that could affect the operation of the robot, and the behavior (or lack of it) that would result. Then, when the inevitable happens, you will have a set of hypotheses about why the robot failed to work. The hypotheses might be wrong, but that is OK: Wrong assumptions lead to right deductions if you are willing to discard assumptions that are not supported by experimentation.

Stiquito Walks in a Straight Line—This is perfect. There is no trouble. Congratulations on a good job!

Stiquito Walks to the Left or Right, But Not Straight—This is OK, and is very common. The reason is that one or more of the legs is crooked, or does not operate at full extension. Check for loose actuator wires; loose crimps; poor electrical connections (look for open or high-resistance paths from the power supply to the legs and back); a weak power supply (nitinol draws about 180 milliamps per leg, which will drain a 9-volt cell after several weeks of occasional use); legs that are not parallel, or ratchet feet bent at different angles.

Stiquito Does Not Walk at All—If there is nothing obviously broken or loose (actuator wires, crimps, control wiring, magnet wire), then check the power source. The 9-volt cell might be dead. Check the power bus and the body crimps for shorts. Check for shorts between connecting crimps (it might be necessary to hold them apart with a piece of tape). If the ratchet feet are not bent, or if Stiquito is operated on a smooth surface, the legs might move but will not catch the surface—and Stiquito will thrash around but not walk. Stiquito walks best on lightly textured surfaces such as indoor-outdoor carpet, a cloth-covered book, pressboard, or poured concrete. Stiquito walks poorly on glass, smooth plastic, or tiled floors.

Leg Moves to Full Extension (4–5 millimeters)—This is perfect. You obviously built Stiquito carefully.

Leg Does Not Move at All—The probable causes include a very loose actuator wire or a dead 9-volt cell. Check the cell with a voltmeter. If the power is OK, then check the electrical connection between the control wire and the body crimp. Use the voltmeter or apply power to various points in the circuit until the leg moves. Do not use a 9-volt cell to test a single leg; it will snap the actuator wire. Test single actuator wires using 3 volts supplied by two 1.5-volt AA cells in series. If the leg cannot be activated from points farther away in the control circuit, then there is an open circuit between that point and the actuator wire. If the electrical connections are good, examine the actuator wire. If it is loose, or the body or leg crimps are not tight, the actuator is too slack to operate. Tighten the actuator wire (it might be necessary to remove the old leg crimp and attach a new one), then test the leg. It should work.

Leg Moves Slightly (1–3 millimeters)—The actuator is probably loose, but is taut enough to take up the slack and then move the leg. Retension the actuator, replacing the leg crimp. Another cause is increased resistance as the crimps age. Aluminum oxide builds up inside the crimps; this can be alleviated by operating the leg, causing the retaining knots to expand and improve contact with the aluminum inside the crimp. Squeezing the crimp also helps to improve the electrical connection.

Leg Moves in One or More Jerks—There is probably an intermittent open or shorted connection. **If the leg jerks backward continuously** in small increments, **remove power immediately** or the actuator wire might be damaged. Check the manual controller for an intermittent connection between the double switch and the contacts.

Leg Heats Up, Smokes, and/or Melts Plastic Near Body Crimp—A little smoke is typical when the actuator wires are first used, probably as oils or oxide on the nitinol wire burn away. But **if there is a lot of smoke** (equivalent to a cigarette left on the edge of an ashtray), or **if something smells hot, like burning or melting plastic, remove power immediately** or the actuator wire might be damaged. There might be a short in the wiring. You might be powering a slack actuator for longer than one second: A slack actuator will not work. Continuing to apply power to the leg will only cause the nitinol actuator wire to overheat, damaging the wire and heating the body crimp enough to melt the plastic body.

Leg Works Well for a While, Then Movement Stops Altogether—The actuator wire has developed slack, a connection has broken, or the battery is dead. If the actuator wire is slack, it might need to be retensioned and recrimped. Check for broken connections, especially where the magnet wire is crimped. Try a fresh 9-volt cell.

Leg Works Well for a While, Then Movement Diminishes by 2 Millimeters or More—The actuator is probably somewhat loose, but is taut enough to take up the slack and then move the leg. Another cause is increased resistance as the crimps age. Apply power to the leg and/or squeeze the body and leg crimps to improve the electrical connection.

Magnet Wires or Control Wires Break—Magnet wire breaks easily. If it does break, remove the connecting crimp, sand away some insulation, and crimp it back into place. If the control wires break, this will probably occur as the hardwired gaits are being changed. Avoid breakage by using the V-grooves to hold the control wire while the body crimp is press-fitted. Do not bend any wires back and forth repeatedly; the wire will become fatigued and break.

Actuator Wire Breaks—If the actuator wire breaks during assembly, it was sanded too much or was nicked (probably while removing the knot from a music-wire leg). Actuator wires can also break from these causes during operation. Do not power a single actuator wire with a 9-volt cell; the wire will contract so rapidly that it cannot overcome the inertia of the leg, and the actuator wire will snap.

Other Controllers—After you have mastered making Stiquito walk with the manual controller, you are ready to use an analog or digital controller. See Chapters 6 through 16 for ideas on how to further control Stiquito.

REFERENCES

1. Dynalloy, Inc. *Technical characteristics of Flexinol® actuator wires.*

2. Gilbertson, R.G. 1992. *Working with shape memory wires.* San Leandro, Calif.: Mondo-Tronics, Inc.

3. Mills, J.W. 1992. Area-efficient implication circuits for very dense Łukasiewicz logic arrays. *Proc. 22nd Int'l Symp. Multiple-Valued Logic.* Sendai, Japan: IEEE Press.

4. Mills, J., and C. Daffinger. 1990. An analog VLSI array processor for classical and connectionist AI. *Proc. Application Specific Array Processors.* Princeton, N.J.: IEEE Press.

5. Brooks, R. 1990. A robot that walks: Emergent behaviors from a carefully evolved network. *Neural Computation* 1, no. 2: 253-262.

6. Beers, R. 1991. An artificial insect. *American Scientist* 79 (Sept.-Oct.): 444-452.

7. Wilson, E.O. 1975. *Sociobiology: The new synthesis.* Harvard University Press.

8. Ballard, D.H., and C.M. Brown. 1982. *Computer Vision.* Englewood Cliffs, N.J.: Prentice-Hall, Inc.

9. See Chapters 6 through 16 for examples.

Chapter 3

Building Stiquito II and Tensipede

Jonathan W. Mills

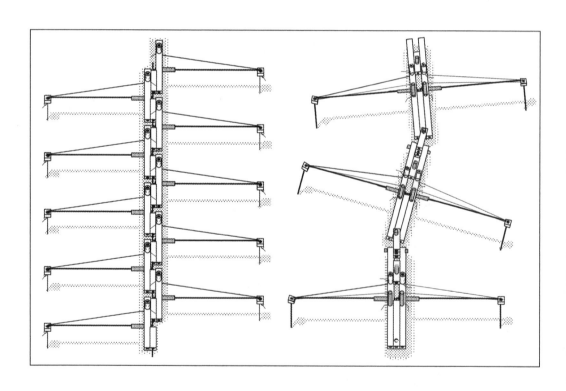

Stiquito II is an improved Stiquito: It is larger, modular, has an articulated body and two-degree-of-freedom legs, and carries more weight. Tensipede is a simple, modular robot centipede. Stiquito II and Tensipede are easier to build than Stiquito. Both are larger and easier to handle. The assembly sequences are simpler and require only common, sturdy tools. A mistake typically affects only one module instead of the entire robot.

This chapter contains three major sections: a tutorial that introduces necessary skills as the reader builds a practice leg; the instructions to build Tensipede; and the instructions to build Stiquito II.

BUILDING STIQUITO II AND TENSIPEDE IS AS EASY AS 1-2-3

1. **Skim through the chapter** to familiarize yourself with its contents.

2. **Build the practice leg** to develop the skills you need to build either robot.

3. **Follow the instructions** to build either Stiquito II or Tensipede.

INTRODUCTION

Look for this helping hand. It will point to a hint to build a better robot.

A Brief History of Stiquito

Legged robots are typically large, complex, and expensive. These factors have limited their use in research and education. Few universities could afford to construct robot centipedes, or 100 six-legged robots to study emergent cooperative behavior; even fewer could give each student in a robotics class his or her own walking robot.

The introduction of Stiquito in 1992 changed that. Stiquito was developed from a larger and more complex robot called Sticky because it looked like an insect commonly called a walking stick. Sticky was very difficult to assemble and rather fragile, so a simpler design was sought. During this process multiple components were combined into single, robust mechanical devices. The best example is Stiquito's leg, which carries the weight of the robot, acts as a leaf spring to provide bias force for the nitinol, and delivers the current that powers the nitinol actuator. A series of five small robots was eventually built that culminated in Stiquito. Stiquito is a small, simple, and inexpensive six-legged robot that has been used as a research platform to study computational sensors, subsumption architectures, neural gait controllers, emergent behavior, and machine vision. Stiquito has also been used to teach science in primary, secondary, and high school curricula.

Stiquito II and Tensipede were developed to provide larger and more robust platforms for research at Indiana University's Analog VLSI and Robotics Laboratory, and yet be easy enough to build that they could be used educationally. Even though both robots are made from simple modules, if you have never built models, assembled electronic circuits, or used hand tools to cut plastic, bend wire, or drill small holes, I urge you to build the practice leg before attempting to build your robot. Care and patience are still needed to assemble a Stiquito II or a Tensipede.

A Comment on Stiquito from the Author

In nature Stiquito and Stiquito II would be greedy, blind, crippled, retarded ants. They consume electricity voraciously; they have no sensors or effectors (at least as described in this chapter); their legs lack the numerous degrees of freedom of those of an ant; their controller, an IBM personal computer, is overwhelmingly large, clunky, and incapable compared to the ganglia that serve an ant (at least to accomplish an ant's tasks).

So why build Stiquitos?

Well, I build them because they are simple, they are fun, and they let me investigate robotics without getting a million-dollar grant. I build Stiquitos because I have yet to construct anything the size of an ant that has the capability of an ant. But I'm working on it.

I encourage all readers of this book to improve on Stiquito, and I assure you that if you build a better ant-like robot, I'll be standing in line to get one.

About This Chapter

The chapter is structured to help the beginning robot builder. After conducting numerous Stiquito workshops, and receiving more than 5,000 electronic mail messages about the robot and the first technical report, the following steps were taken to improve the presentation.

- Large figures that illustrate the desired appearance of each major component are presented at the beginning of the section that describes how the component is built. This helps the builder understand the steps in the assembly process and visualize the relationships between the parts and the finished component. The illustration of the component is repeated at reduced size to mark the end of the assembly. This helps break down the assembly process into related groups of steps.

- *Cues*—small pictures—are used to focus the builder's attention during each step of the assembly. Cues warn the builder in advance of common mistakes, point out critical steps in the assembly, identify which parts must be made precisely, or indicate when the assembly can tolerate a little sloppiness. Cues are my attempt to share my robot-building experience with the reader.

- The names of the parts have been chosen to emphasize the modularity of the design. While there is no need for complex part numberings—after all, Stiquito II is not a space shuttle—there is still a need to avoid cumbersome and lengthy descriptions of a specific (and usually very small) part. All parts are made during the first steps in the assembly process. This seems to reduce stress among first-time builders by breaking the apparently complicated assembly sequence into simple, tangible parts.

COSTS, PARTS, AND MATERIALS

Materials alone for a single robot cost about $17. Nitinol wire is available in a minimum length of 1 meter for $10. It is the most expensive part of the robot. The prorated, 1996 cost of materials used to construct Stiquito II in volume (500 or more) is less is $10.96 each.

Amount	Cost	Part number	Item
30 in.	$1.00	Plastruct ST-4	$1/8''$ square ABS plastic rod
24 in.	$0.50	K&S Engineering No. 117	$1/16''$ outside diameter copper tube
12 in.	$0.30	K&S Engineering No. 102	$1/8''$ outside diameter aluminum tube
36 in.	$0.15	K&S Engineering No. 501	.032'' music wire
3 in.	$0.01	n/a	28 ga copper wire (telephone wire)
1,500 mm	$9.00	Dynalloy .004'' dia. 70°C	.004'' (100μm) nitinol wire (Flexinol)

The plastic stock is available from Plastruct, Inc., and the aluminum tubing and music wire are available from K&S Engineering. Hobby shops carry these products. Flexinol can be ordered in lengths of one meter or more from Dynalloy, Inc. or from MondoTronics, Inc. Refer to the appendix for addresses of these companies.

TOOLS AND RULES

Tools

The following tools are **required** to build Stiquito II.

- Needlenose pliers
- Wire cutters
- Rubber bands (used to clamp parts and grip objects with pliers without holding them)
- Small hobby knife (X-Acto type)
- Pin vise with $1/16''$ drill bit
- 320-grit fine sandpaper
- Model cement (methyl ethyl ketone; also contained in waterbed patch cement)
- Two AA batteries and holder

The following tools are **optional**.

- Locking forceps (hemostats)
- Ruler marked in millimeters (optional because illustrations are drawn life-size)
- C-clamps (instead of rubber bands)
- Dremel Moto-Tool with $1/16''$ drill bit

Ground Rules

Look for these cues (small pictures) as you follow the instructions to build your robot. They will warn you in advance of mistakes you should avoid, point out critical steps, identify which parts must be made precisely, or tell you when a little sloppiness is OK.

Cue	Meaning
✓	**Do** take the action indicated. No particular precision is required.
✗	**Don't** do this! It is a mistake.
▲	**Apply plastic cement** to the part surfaces marked with a heavy dotted line (▬ ▬ ▬ ▬ ▬).
|||	**Work as carefully as you can.** The dimensions are critical. Stay within 0.5 millimeters or less of the required measurement.
⟩⟩⟩	**Some variation is allowed.** The dimensions are not critical. Stay within one or two millimeters of the specified measurement.
●	**A tight fit is necessary.** A loose or sloppy fit will prevent your robot from working well, and may keep it from working at all.
◉	**A loose or sloppy fit is OK.** Your robot will work fine if the fit is not tight.

PRECAUTIONS

✓ Always follow the manufacturer's instructions when using a tool.

✓ Wear safety glasses to avoid injury to your eyes from broken tools, or pieces of plastic or metal that may fly away at high velocity as a result of cutting or sawing. Be especially careful when cutting music wire with wire cutters (the cut wire can be forced into your finger, for example).

✓ Use motorized tools such as Dremel Moto-Tools carefully. Motorized tools are not needed to construct Stiquito, although they are helpful.

✓ Use a piece of pressboard, dense cardboard, or wood to protect your work surface.

✗ **Avoid inhaling the fumes of model cement.**

✓ Work in a well-ventilated area.

PRACTICE LEG

 You can **improve the quality of your robot** by building this practice leg first.

This section introduces the skills to build either robot as you build a nitinol-actuated leg.

This is what the completed leg will look like from the front and the top.

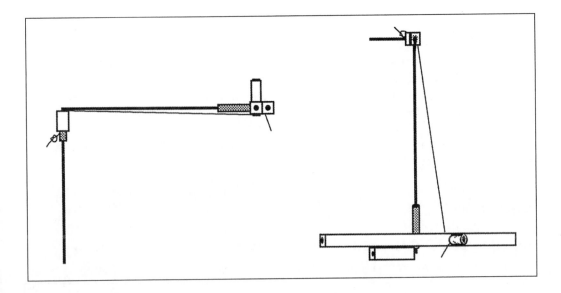

Let's begin now.

 Keep track of your progress by **crossing off each step as you complete it**. Mark through the little picture to the left of each step with a highlighter or colored marker to show that you have finished it.

Measure and Cut Parts

The first two skills you will learn are used to make parts from the raw materials provided in the kit.

Measuring—A ruler is optional because critical illustrations that provide dimensions are drawn to scale. Use these illustrations to measure your parts, locate holes, and so on.

Measure twice, cut once to prevent most mistakes. If you choose not to use the illustrations, use any metric ruler marked in millimeters. Look directly down at the ruler so that your measurement is not affected by your viewing angle. Mark the location by scoring the tube or rod lightly with the knife.

Cutting Plastic Rod and Metal Tubing—Before cutting, check that your fingers are not in the way of your knife, and that a slip of the knife will not damage anything nearby. Direct the knife away from yourself to avoid injury. Cut tubing and plastic rod by scoring it firmly around the outside of the tube or rod. Score plastic rod by rocking the knife back and forth on each side, working your way around all four sides. Score tubing by slowly rolling the tubing under the knife while pressing down on it firmly. After scoring the tube or rod, bend it back and forth gently to snap it at the score.

Cutting Music Wire—Cut music wire with wire cutters. Use heavy-duty wire cutters if you have them; otherwise cut the wire near the bottom of the cutters (**not** at the tip, where the wire cutters are more easily damaged). Be careful, as the cut wire may shoot away from the cutters. Hold the wire along its length, **not** by its end, or the cut wire may be forced into your finger.

Now measure and cut the following parts for each assembly. The cues in the part column tell you how carefully to make each part.

Leg Segment

Quantity	Length	Material	Part	Illustration
1	50 mm	$1/8$" ST-4 plastic rod	§§§ leg rod	(see below)
1	10 mm	$1/8$" ST-4 plastic rod	§§§ leg holder	
1	100 mm	.32" steel music wire	§§§ leg	(see below)
1	10 mm	$1/16$″ o.d. copper tube	§§§ leg bushing	

Actuator

Quantity	Length	Material	Part	Illustration
1	75 mm	.004″ 70° C nitinol wire	||| actuator wire	(see below)
1	5 mm	$1/8$" ST-4 plastic rod	||| knee clamp holder	
2	8 mm	$1/16$" o.d. copper tube	||| actuator clamp	
1	5 mm	$1/8$" o.d. aluminum tube	||| outer clamp	

Prepare the Leg Rod

Drilling and deburring are needed to make holes to hold actuators and legs.

Drilling—Mark the location of holes by lightly scoring the plastic body segment. Start the hole by twisting the point of the knife blade back and forth at the location of the hole. This makes a shallow hole to hold the drill bit in place. Brace the part on a sheet of scrap wood or plastic to protect the work surface. Hold the pin vise perpendicular to the part when drilling, and twist the bit back and forth with a short but firm motion, applying moderate downward pressure. If the drill bit sticks, pull it out then begin again. Be careful not to drill through the material used to brace the part. After making the hole, twist the bit as you slide it into and out of the hole. This will ensure that the hole is the correct diameter, cylindrical, and smooth inside. Blow into the hole to remove any scraps of material left inside.

Deburring—Cutting and drilling can leave rough edges, or burrs, on some parts. Remove the burrs by sanding the rough edge with fine sandpaper (320 grit), by trimming the burr with a small knife, or by lightly abrading the part with a drill bit held in a pin vise. Leaving burrs on parts, especially clamps, can cause the nitinol actuators to break. Parts that are press-fitted may bend or break during assembly if not deburred.

Now prepare the leg rod.

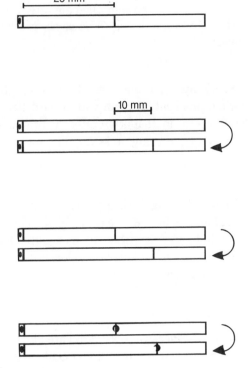

§§§   Measure 25 millimeters from one end of the leg rod. This is the location of the leg hole.

✓   Mark the location of the leg hole by lightly scoring it with the knife. This also defines the top of the leg rod.

||| Measure 10 millimeters further to the right of the center of the leg rod, then turn the rod on its side. The actuator hole is located here, on the side of the leg rod.

✓   Mark the location of the actuator hole by lightly scoring the leg rod with the knife.

✓   Start the leg hole and actuator hole by twisting the tip of the knife back and forth in the middle of the leg rod at each score.

||| Each hole should be centered in the width of the leg rod (the middle of the score).

||| Drill the leg hole and the actuator hole.

✓   Debur and clean out any scrap by blowing through the holes and running the drill bit in and out through the holes.

Bend the Leg

Bending music wire makes joints that retain the leg and the actuator wires.

Bending—All bends in the music wire are 90-degree angles. Use a pair of needlenose pliers and hold the music wire near the joint of the pliers (at the thick part) to get a good grip as you make the bend. Don't straighten out the wire if a bend isn't quite perfect. Adjust the bend by twisting it. Leave it alone if it is almost at 90 degrees unless a later step in assembly requires a better bend.

Now bend the leg.

||| Bend the leg in a sharp 90-degree bend...

§§§   ...about 5 millimeters from one end.

90°

5 mm

Attach the Leg and Leg Holder to the Leg Rod

Tubing must be smoothed out after it is cut and before it is inserted into the plastic rod. This allows the tubing to be easily press-fitted without enlarging the hole into which it is inserted.

Swaging—Roll the ends of cut tubing back and forth over the top of the work surface while pressing on them with a hard flat object, like the back of a metal ruler. This will remove the lip that is left after the tubing has been scored and snapped apart, and that prevents it from being press-fitted easily.

Press Fitting Tubing into Plastic Rod—Push firmly but carefully to insert the tubing into plastic rod. Press-fitting does not require excessive pressure. If you push hard and nothing happens, debur the part and hole, and swage the tubing again. The tubing should not fall out of the hole holding it; if it does, the hole is too large. Discard the part with the hole and make a new one.

Gluing—Pieces of plastic rod are glued together to fabricate larger components like body segments and c-joints. Model cement is used to bond plastic parts to each other. Do not use too much cement or the plastic may melt. Let the plastic get tacky before pressing parts together. Hold parts together for 15 seconds to bond them, then clamp the assembly and let it dry at least 2 hours before cutting or drilling it.

Now attach the leg and leg holder to the leg rod.

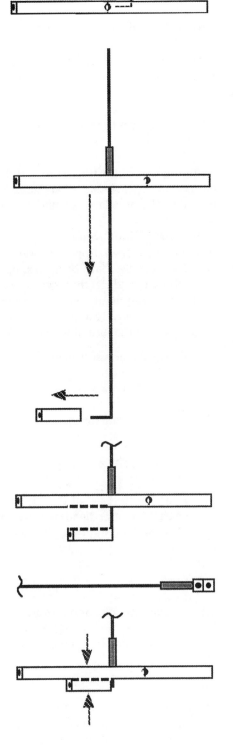

✓ Swage the leg bushing.

● Insert the leg bushing into the leg hole.

◉ Insert the long length of the leg into the leg bushing.

◉ Insert the short 5 millimeters length of the leg into the hole in either end of the leg holder.

✓ Push the leg rod along the leg until it touches the leg holder.

✗ Do **not** continue if the leg rod and the leg holder aren't parallel. The bend in the leg is not 90 degrees. Pull the leg out of the leg rod and the leg holder, adjust the bend, then insert the leg and try again.

§§§ Pull the leg rod and the leg holder apart about 10 millimeters, and...

◭ ...apply plastic cement to the inner face of the leg rod and the leg holder.

✓ Let the glue get tacky, then push the leg rod and the leg holder back together.

||| Align the leg rod with the leg holder so they are in the same plane.

◭ Lightly apply some plastic cement along the joints between the body segment sides and the leg pair holder. No gap should be visible. The joint should be even and smooth.

✓ Clamp the leg assembly.

Let the leg assembly dry for **at least** 2 hours. When you make the leg assembly you can get away with waiting just 15 minutes, but don't make a habit of it.

Attach the Nitinol Actuator

The parts of the nitinol actuator must fit tightly without any play, or the nitinol will waste most of its force taking up the slack due to sloppy assembly. Take extra care during these steps.

Attaching Actuator Clamp to Plastic Rod—This is the same as press-fitting tubing into the plastic rod.

Reaming Out the Knee Clamp Holder—Reaming is the same as drilling, but is used to enlarge an existing hole. Reaming is used to enlarge the hole in the center of the ST-4 plastic rod so that it is large enough to hold the actuator clamp.

Clamping Nitinol—Smooth and swage the ends of the copper actuator clamps to avoid breaking the nitinol. Apply tension to the nitinol using locking forceps or needlenose pliers whose handles have been wrapped tightly with several rubber bands. Keep the nitinol taut until it is held tightly between the outer wall of the actuator holder and the inner side of either the aluminum outer clamp or the plastic knee clamp holder. You can remove the outer clamp and retighten the nitinol if necessary.

Now attach the nitinol actuator to the leg rod and the leg.

✓   Swage the actuator clamp.

●   Insert the actuator clamp into the actuator hole.

||| Bend the free end of the leg down in a sharp 90-degree bend, toward the side of the leg rod **opposite** the actuator clamp...

§§§   ...about 50 millimeters from the leg rod. This will leave about 40 millimeters from the knee to the foot.

||| Twist the free end of the leg right or left until it makes a 90-degree angle with the leg rod when viewed from the side.

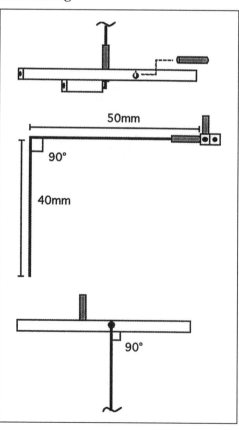

§§§ Use the drill to ream out (enlarge) the hole that runs through the center of the knee clamp holder.

✓ Grasp the knee clamp holder with the needlenose pliers so that the drill bit does not slip.

✓ Ream the entire length of the hole.

§§§ Tie an overhand knot in the nitinol actuator and pull it finger-tight so it is about 3 millimeters from one end of the nitinol actuator.

◉ Insert the free end of the actuator into the knee clamp holder and slide the knee clamp holder to the knot.

✓ Swage the actuator clamp.

● Press the actuator clamp into the knee clamp holder at the hole nearest the knot.

✓ Push it all the way through the knee clamp holder. The actuator clamp will protrude from the end of the knee clamp holder.

✗ Do **not** let the nitinol actuator cross over the end of the actuator clamp, or it will be cut off.

✗ Do **not** use excessive force to press-fit the actuator clamp, or the nitinol will break now or after a few activation cycles. If it is difficult to press-fit the actuator clamp into the knee clamp holder, ream out the knee clamp holder again and swage the ends of the actuator clamp.

◉ Slide the knee clamp assembly onto the end of the leg. The free end of the nitinol actuator must exit the knee clamp assembly facing **toward** the actuator clamp.

◉ Now thread the free end of the nitinol actuator through the flush end of the actuator clamp and out its top.

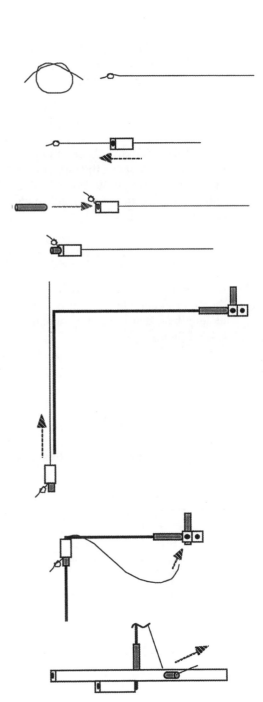

✓ Grip the free end of the actuator with the locking forceps and let it hang freely to provide tension to the actuator.

● Press-fit the aluminum outer clamp over the nitinol and the actuator clamp. It will be a snug fit.

✗ Do **not** use excessive force to press-fit the outer clamp, or the nitinol will break. If it is difficult to press-fit the outer clamp, ream out the inside ends slightly with the knife.

⌇⌇⌇ Use the wire cutters to **gently** crimp the outer clamp about halfway down its length so that the nitinol actuator is held tightly.

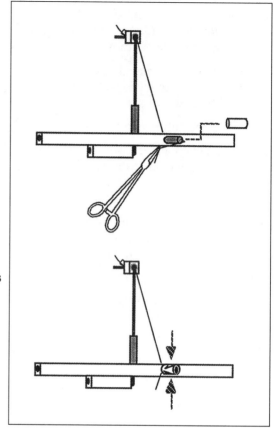

The completed leg will look like this from the front and the top (reduced size illustration).

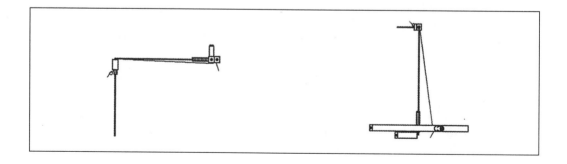

Apply 3 volts from the two AA cells at the points indicated, and the leg should bend toward the actuator clamp.

✗ Do **not** use more than 3 volts, or the nitinol actuator will snap.

✗ Do **not** apply power to two points close together on the nitinol wire, or the nitinol actuator will overheat and melt or break.

✗ Do **not** touch a heated nitinol wire. You could burn yourself.

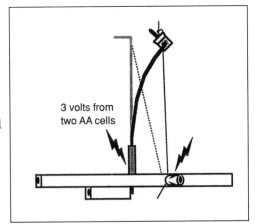

3 volts from two AA cells

 Congratulations! You have successfully built the practice leg. **Take a break,** then you can continue by building either Stiquito II or Tensipede.

TENSIPEDE: A ROBOT CENTIPEDE BASED ON THE PRACTICE LEG

The materials and quantities specified earlier in this chapter allow you to **build either Tensipede or Stiquito II, but not both.** You must choose which robot you want to build, or obtain more materials to build both.

You can build a number of interesting robots based on the practice leg. As an example, here is the design for Tensipede, a 10-legged robot centipede that high-school junior John Dankanich of Hammond, Indiana, assembled and tested during his summer internship at Indiana University's Analog VLSI and Robotics Laboratory.

If you work carefully, your finished Tensipede will look like this:

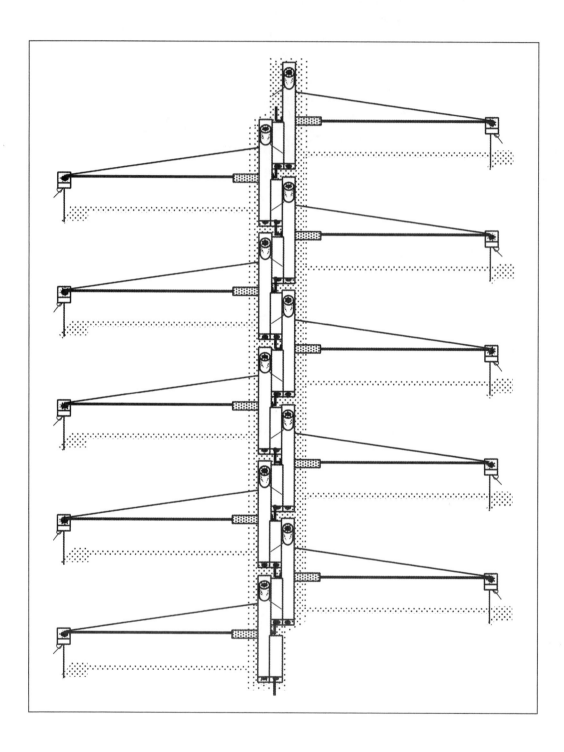

Start by making the required number of parts to build 10 legs that use a shorter leg rod.

LEG SEGMENT

Quantity	Length	Material	Part	Illustration
10	25 mm	$1/8$" ST-4 plastic rod	||| leg rod	
10	10 mm	$1/8$" ST-4 plastic rod	||| leg holder	
10	100 mm	.032" steel music wire	$$$ leg	(see below)
10	10 mm	$1/16$" o.d. copper tube	$$$ leg bushing	

ACTUATOR

Quantity	Length	Material	Part	Illustration
10	75 mm	.004" 70 C nitinol wire	||| actuator wire	(see below)
10	5 mm	$1/8$" ST-4 plastic rod	||| knee clamp holder	
20	8 mm	$1/16$" o.d. copper tube	||| actuator clamp	
10	5 mm	$1/8$" o.d. aluminum tube	||| outer clamp	

SPINE

Quantity	Length	Material	Part	Illustration
1	150 mm	.032" steel music wire	$$$ spine	(see below)

MAKE FIVE LEGS

Note the changes: The leg rod is shorter, and the leg holder must be reamed out to hold the spine.

||| Measure 12 millimeters from one end of the leg rod. This is the location of the leg hole.

✓ Mark the location of the leg hole by lightly scoring it with the knife.

||| Measure 10 millimeters farther to the right on the leg rod, then turn the rod on its side. The actuator hole is located on the side of the leg rod.

✓ Mark the location of the actuator hole by lightly scoring the leg rod with the knife.

✓ Start the leg hole and actuator hole by twisting the tip of the knife back and forth in the middle of the leg rod at each score.

||| Each hole should be centered in the width of the leg rod (the middle of the score).

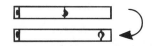

||| Drill the leg hole and the actuator hole.

$$$ Use the drill to ream out (enlarge) the hole that runs through the center of the leg holder.

✓ Debur and clean out any scrap.

✓ Make five (5) legs with the shorter 25-millimeter leg rod.

✓ **All other parts are the same as for the practice leg.**

✓ **Follow the instructions to make the practice leg.**

✓ When you are through, your five legs should look like this from the top.

Make Five Mirror-Image Legs

✓ Make five (5) legs with the shorter leg rod that are **mirror images** of the first five.

✓ When you are through, your five mirror-image legs should look like this from the top.

✗ Do **not** make them identical to the first five legs or your Tensipede will only be able to walk in circles (unless you do some tricky programming).

Attach Legs to a Spine

◉ Thread the legs onto the spine
through the leg holders, alternately attaching a leg and a mir-
ror-image leg until all legs are
attached to the spine.

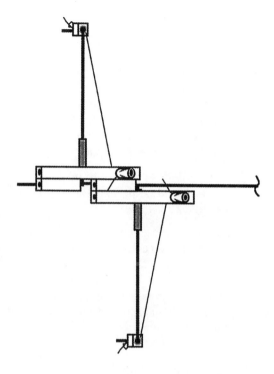

When complete, your Tensipede should look like this (reduced size illustration).

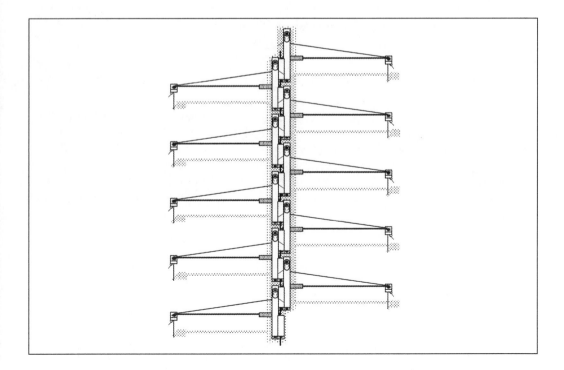

Finish by Wiring the Tensipede to the Controller

✓   Build the PC interface described in Chapter 6.

✓   Wire the Tensipede to the controller. Attach the spine to the positive (leg+) power source. The actuator clamps on the leg rods are attached to the Darlington drivers' outputs.

✓   Write a BASIC program to make the Tensipede walk (see the example in Chapter 6).

✗   Do **not** activate more than three legs on each side simultaneously. The Darlington drivers can sink about 700 milliamps, enough to drive only three legs on each side at once. If more are simultaneously activated, the legs will not move to their full extent.

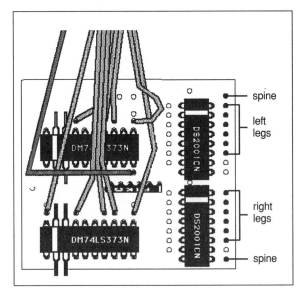

    Tensipede can carry a large payload. You may want to put a single chip controller, like a PIC 1600 or a BASIC Stamp on the robot, along with batteries and driver circuits for the nitinol actuators. **Tensipede, however, doesn't turn very well.** You may want to try designing an articulated joint in the middle of the robot to allow it to turn more easily. For example, a small spring could be used to connect a spine made in two pieces.

STIQUITO II: A HEXAPOD WITH A MODULAR ARTICULATED BODY

Stiquito II is a larger Stiquito with an articulated body that is designed to move freely, allowing the robot to walk up and down shallow stairs, climb small inclines, negotiate rough terrain, and turn to avoid obstacles, all without explicitly being programmed to do so.

Read the instructions through completely before you start to assemble Stiquito II. Check that you have all the materials and tools needed before you start. Plan to spend about four to eight hours (or an equal amount of time in several shorter sessions) building Stiquito II.

If you work carefully, your finished Stiquito II will look like this.

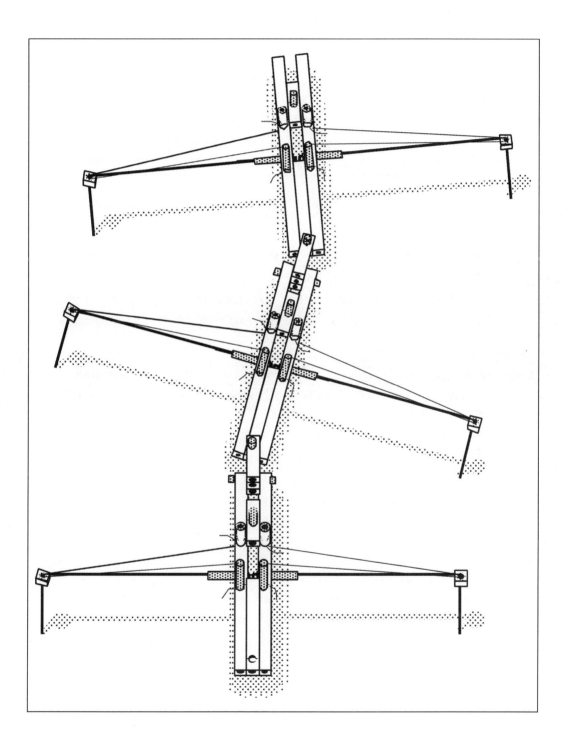

Measure and Cut Parts

Measure and cut the following parts for each assembly from the materials indicated.

Body Segment

Quantity	Length	Material	Part	Illustration
6	50 mm	1/8" ST-4 plastic rod	||| side	
3	23 mm	1/8" ST-4 plastic rod	||| leg pair holder	
3	10 mm	1/8" ST-4 plastic rod	||| power bus holder	
6	100 mm	.032" steel music wire	SSS leg	(see below)
6	10 mm	1/16" o.d. copper tube	SSS leg bushing	

C-Joint

Quantity	Length	Material	Part	Illustration
4	10 mm	1/8" ST-4 plastic rod	||| top/bottom side	
2	5 mm	1/8" ST-4 plastic rod	||| inner side	
4	15 mm	1/16" o.d. copper tube	SSS axle	

Actuators

Quantity	Length	Material	Part	Illustration
6	150 mm	.004" 70 C nitinol wire	||| actuator wire	(see below *not to scale*)
6	5 mm	1/8" ST-4 plastic rod	||| knee clamp holder	
6	8 mm	1/16" o.d. copper tube	||| knee clamp	
6	8 mm	1/16" o.d. copper tube	||| horizontal clamp	
6	15 mm	1/16" o.d. copper tube	||| vertical clamp	
6	5 mm	1/8" o.d. aluminum tube	||| outer clamp	

Power Bus

Quantity	Length	Material	Part	Illustration
3	8 mm	1/16" o.d. copper tube	SSS receptacle	
3	25 mm	28 ga copper wire	SSS wire	

MAKE THREE BODY SEGMENTS

A body segment looks like this (shown slightly reduced in size so that both legs can be seen).

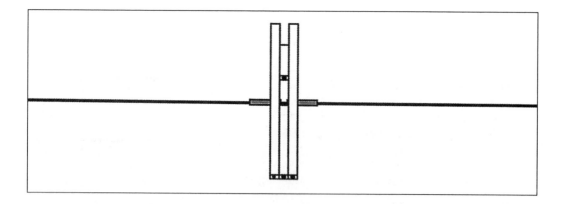

Make Six Body Segment Sides

||| Measure 25 millimeters to the center of each body segment side.

✓   Mark the center by lightly scoring it with the knife.

✓   Start the leg bushing holes by twisting the tip of the knife back and forth in the middle of the body segment side at the score.

||| The hole should be centered in the width of the body segment side (the middle of the score).

||| Drill the leg bushing holes.

✓   Debur and clean out scrap.

⬤   Insert the leg bushings into the leg bushing holes.

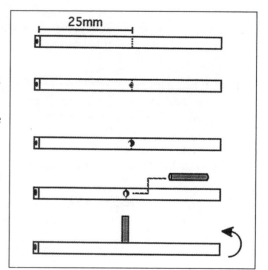

Make Three Leg Pairs

||| Enlarge the hole in one end of each leg pair holder by grasping the leg pair holder with the pliers and reaming (drilling out) the hole with the 1/16″ drill bit...

〰〰〰   ...to a depth of about 1/2″.

||| Bend each leg in a sharp 90-degree bend...

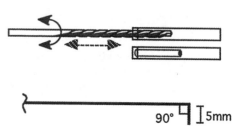

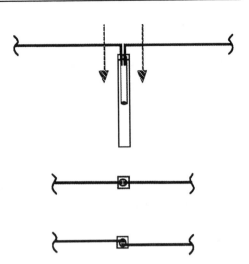

$$$ ....about 5 millimeters from one end.

◉   Insert the short 5 millimeters length of
    two legs into the enlarged hole in the
    leg pair holder.

||| Adjust the legs until they are aligned
      along the diameter of the enlarged hole
      in the leg pair holder, viewed from the
      end.

✗   Do **not** allow the legs to get off-center.
    If this happens you will not be able to
    align the body segment with the leg pair
    holder, preventing it from being glued
    securely.

Attach Two Body Segment Sides to Each Leg Pair

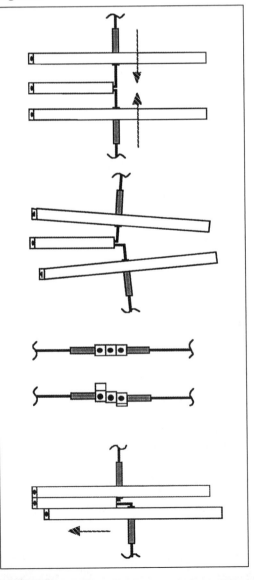

◉   Insert the long length of the legs into
    the leg bushings.

✓   Push the body segment sides along the
    legs until they touch the leg pair holder.

✗   Do **not** continue if the body segment
    sides and the leg pair holder aren't
    parallel. The bend in one or both legs
    is not 90 degrees. Pull the legs out of
    the leg pair holder, adjust the bend,
    then insert the legs and try again.

||| Align the body segment sides with the
      leg pair holder so they are in the same
      plane.

✗   Do **not** continue if the body segment
    sides and the leg pair holder won't line
    up in the same plane. The legs are mis-
    aligned in the enlarged hole in the leg
    pair holder. Twist the leg pair holder
    until the legs are opposite each other.

||| Align the legs so they are opposite
      each other. You may need to slide one
      or both of the body segment sides
      forward or backward.

| | | Align the body segment sides and the leg pair holder so their ends are even. You may need to slide the leg pair holder forward or backward.

✓ When aligned, the body segment should look like this.

§§§ Pull the body segment sides apart slightly, then grasp the body segment by the body segment sides...

✓ ...and rotate the leg pair holder upward and outward.

◭ Continuing to hold the body segment, apply plastic cement to the inner face of each body segment side and the outer two faces of the leg pair holder.

| | | Let the glue get tacky, then rotate the leg pair holder back into alignment with the body segment sides and push the body segment sides against the leg pair holder.

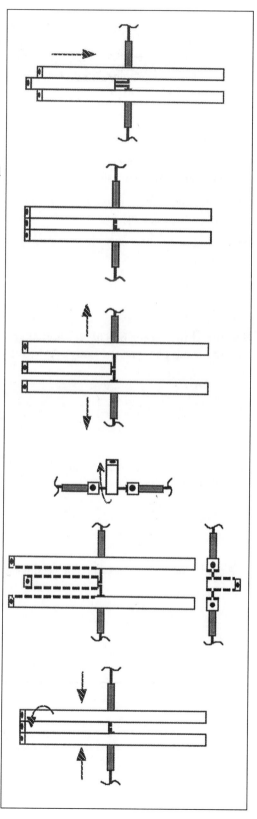

Lightly apply some plastic cement along
the joints between the body segment sides
and the leg pair holder. No gap should be
visible. The joint should be even and smooth.

Apply plastic cement...

...to 10 millimeters of each inner face of
each body segment side...

...starting 8 millimeters back from the center
of the leg bushing...

...and to the two outer faces of the power
bus holder.

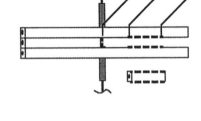

Insert the power bus holder and align it in
the same plane as the body segment sides.

✓ Clamp the body segment just forward
of the leg bushings.

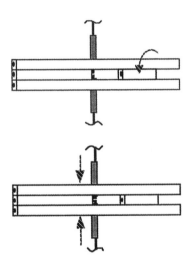

A body segment looks like this (reduced size illustration).

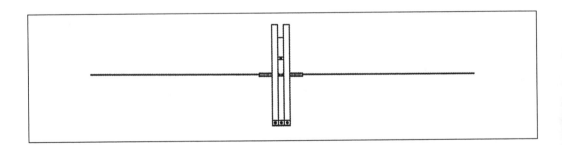

Let the body segment dry for **at least** two hours. Build the remaining body segments if you have not yet done so, or continue to the next section and assemble the c-joints.

Assemble Two C-Joints

💧 Apply plastic cement...

〰 ...to 5 millimeters at one end of each inner face of the c-joint top and bottom sides.

💧 Apply plastic cement to the two outer faces of the c-joint inner side.

✓ Make the c-joint by attaching the top and bottom sides to the c-joint inner side.

||| Align the c-joint top and bottom sides with the c-joint inner side so all lie in the same plane.

||| Align the c-joint top and bottom sides with the c-joint inner side so their ends are even.

✗ Do **not** continue if the c-joint is crooked. Press the c-joint against a flat surface to align it. Break it apart and reglue it if necessary.

💧 Lightly apply some plastic cement along the joints between the c-joint top and bottom sides and the c-joint inner side. The joint should be even and smooth.

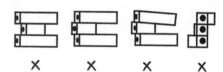

✗   ✗   ✗   ✗

✓ Clamp the c-joint.

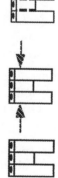

A c-joint looks like this.

 Assemble the other c-joint, then let both c-joints dry for **at least** two hours. It is a good time to **take a break** if you have completed the body segments and c-joints, and are waiting for them to dry.

Drill Holes and Bend Legs on Each Body Segment

 Do **not** continue until the glue holding the body segments and c-joints is dry. If you drill holes in the body segments or c-joints before the glue is dry, you will probably pull their component plastic rods apart.

A body segment with actuator and axle holes drilled in it, and the legs bent downward, should look like this viewed from the top front.

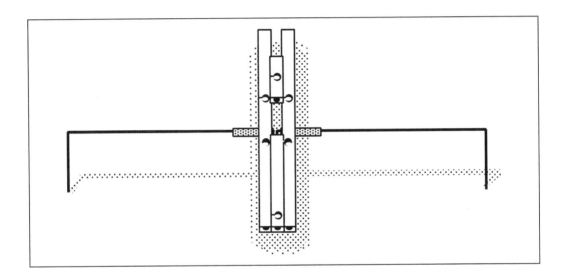

Drill Four Actuator Clamp Holes in Each Body Segment

||| Center all holes along the width of the ST-4 plastic rod.

||| Locate the vertical clamp holes 3 millimeters in front of the center of the leg pair bushing. Start one hole in each body segment side.

||| Locate the horizontal clamp holes 8 millimeters in back of the center of the leg pair bushing. Start one hole in each body segment side.

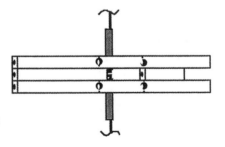

||| Carefully drill these holes. They must be vertical, cylindrical, and ¹/₁₆″ in diameter.

✗ Do **not** make the holes larger than ¹/₁₆″ or the actuators and power bus receptacle will not fit tightly (the actuators are press-fitted ●). It is very easy to make the holes too large, crooked, conical, elliptical, or otherwise non-cylindrical if a motorized drill is used carelessly.

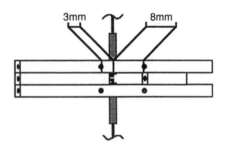

✓ Debur and clean out scrap by blowing through the hole.

Drill One Power Bus Receptacle Hole in Each Body Segment

§§§   Locate the power bus hole about 5 millimeters from the end of the body segment power bus holder (in the middle).

||| Center the hole along the width of the ST-4 plastic rod.

||| Carefully drill the hole. It must be vertical, cylindrical, and $1/16''$ in diameter.

✗   Do **not** make the hole larger than $1/16''$ or the power bus receptacle will not fit tightly.

✓   Debur and clean out scrap.

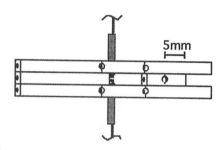

Drill One Vertical Axle Hole in Each Body Segment

||| Locate the body segment vertical axle hole 2.5 millimeters from the end of the body segment farthest from the power bus holder.

||| Center the hole along the width of the ST-4 plastic rod.

✓   Use the knife to start the hole in the leg pair holder.

||| Carefully drill the vertical axle hole.

✓   Debur and clean out scrap.

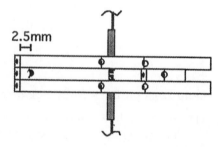

Drill One Horizontal Axle Hole on Each Side of Each Body Segment

||| Trim the ends of the body segment sides nearest the power bus holder so they are even.

||| Locate the horizontal axle holes at the rear of the body segment. Measure 3 millimeters from the end of each body segment side and mark with a score on the **side,** not the top, of **each** body segment side.

||| Center each hole along the width of the ST-4 plastic rod.

✓   Start each hole with the knife.

||| Carefully drill each horizontal axle hole.

✓   Debur and clean out scrap.

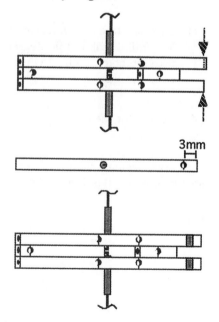

✓   Test the holes by inserting a drill bit through both of them, and rotating it. If you cannot do this easily, then...

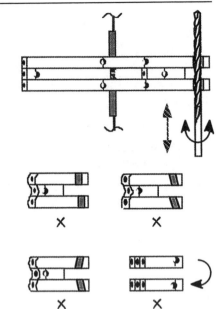

✗   ...do **not** use the body segment because the holes are not **opposite each other** or do not lie **on the same axis.** The axle will be difficult or impossible to insert, or will bind if inserted and prevent the joint from moving smoothly and freely. Discard it and make another.

Bend Legs on Each Body Segment

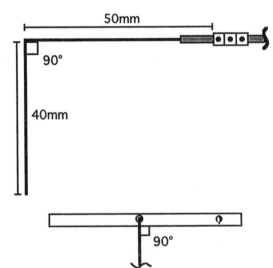

||| Bend the free end of either the right or left leg down in a sharp 90-degree bend...

$$$ ... about 50 millimeters from the leg rod. This will leave about 40 milli-meters from the knee to the foot.

||| Twist the free end of the leg right or left until it makes a 90-degree angle with the body segment when viewed from the side.

✓   Repeat for the other leg.

A body segment with actuator and axle holes drilled in it, and the legs bent downward, should look like this viewed from the top front (reduced size illustration).

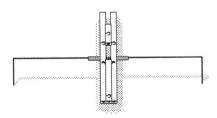

Body segments are **interchangeable.** If you have had trouble drilling the axle holes, you may be able to switch body segments to get a configuration that will work.

Drill Axle Holes in Each C-Joint

Drill Two Vertical Axles Holes in Each C-Joint

||| Trim the ends of the c-joint top/bottom sides at the open end of the c-joint so they are even.

||| Locate the c-joint vertical axle holes at the open end of the c-joint 2 millimeters from the end of each c-joint top/bottom side.

2mm

||| Center all holes along the width of the ST-4 plastic rod.

||| Carefully drill each vertical axle hole.

✓ Debur and clean out scrap.

✓ Test the holes by inserting a drill bit through both of them, and rotating it. If you cannot do this easily, then...

✗   X   X

✗ ...do **not** use the c-joint because the holes are not **opposite each other** or do not lie **on the same axis**. Discard it and make another.

X   X

Drill One Horizontal Axle Hole in Each C-Joint

||| Locate the c-joint horizontal axle hole 2.5 milli-meters from the closed end of the c-joint.

||| Center the hole along the width of the c-joint inner side, then use the knife to start the hole.

||| Carefully drill the c-joint horizontal axle hole.

✓ Debur and clean out scrap.

2.5mm

The finished c-joints look like this.

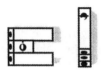

 Now is a good time to **take a break**.

Attach Actuators and Power Bus to Each Body Segment

A finished body segment will look like this when the actuators and the power bus have been attached.

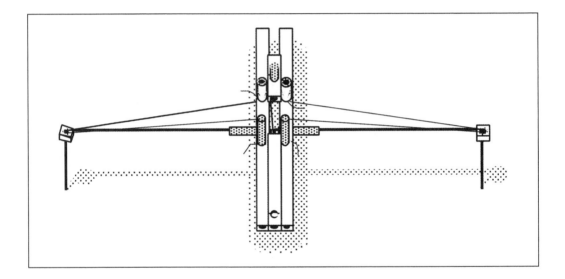

Attach Two Horizontal Clamps to Each Body Segment

✓ Swage the horizontal clamps.

● Insert each horizontal clamp into the horizontal clamp hole until it is flush with the bottom of the body segment.

$$ If the horizontal clamp is loose, remove it and crimp it gently approximately midway along the section that fits into the body segment. This will usually provide enough pressure to ensure a tight fit.

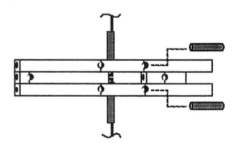

Attach Two Vertical Clamps to Each Body Segment

✓ Swage the vertical clamps.

● Insert each vertical clamp into the vertical clamp hole...

III ...until it protrudes 5 millimeters from the bottom of the body segment.

§§§ If the vertical clamp is loose, remove it and crimp it gently midway along the section that fits into the body segment. This will usually provide enough pressure to ensure a tight fit.

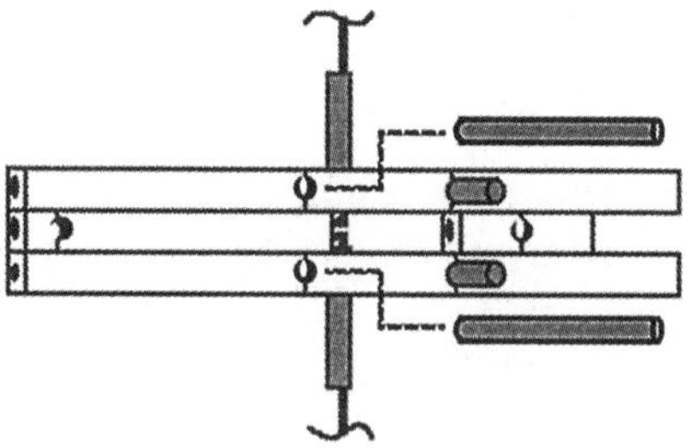

☞ Do **not** use the body segment if the horizontal or vertical clamps cannot be made to fit tightly. The nitinol actuators will work poorly, or not at all. Discard the body segment and make another, being careful to drill the holes exactly 1/16″ in diameter. If you used a motorized drill the first time, try using a pin vise to hold the bit, and drill the holes manually. **Take care that the holes are not too large, crooked, conical, elliptical, or otherwise noncylindrical.**

Attach One Power Bus to Each Body Segment

✓ Swage the power bus receptacle.

● Insert the power bus receptacle into the power bus receptacle hole...

§§§ ...until it protrudes about 5 millimeters from the bottom of the body segment.

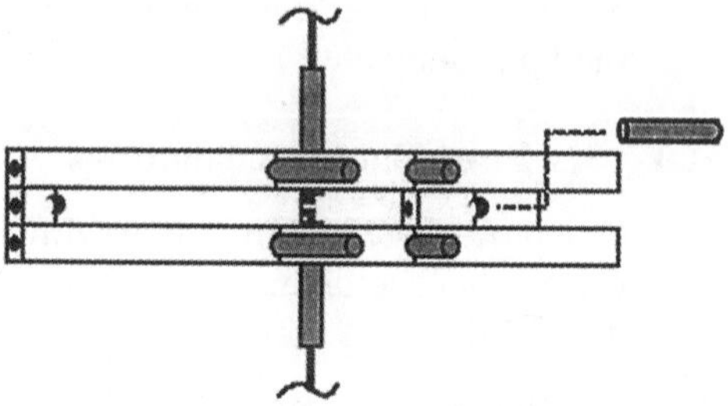

✓ Turn the body segment over.

✓ Strip the insulation from the power bus wire.

● Force the power bus wire into the leg pair holder...

§§§ ...about 10 millimeters. It must be a tight fit to make good electrical contact with the legs.

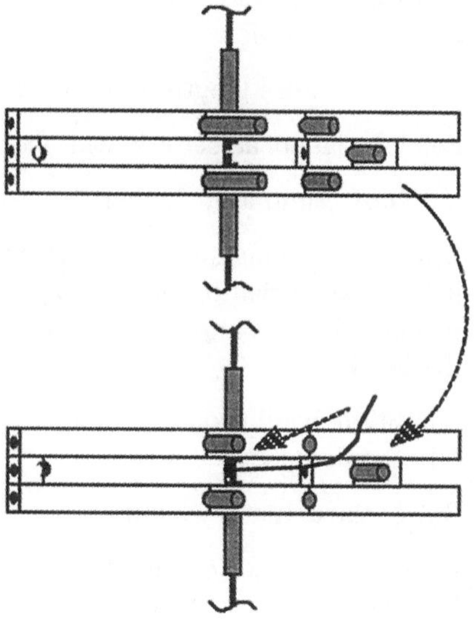

§§§ Bend the power bus wire up about 4 millimeters.

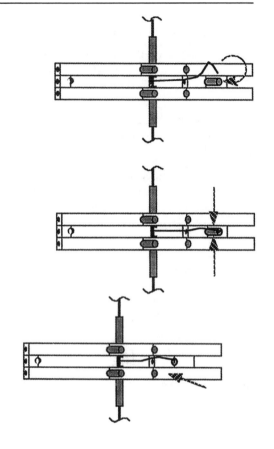

◉ Insert the bent length of the power bus wire into the power bus receptacle.

✓ Firmly crimp the power bus receptacle to make a good electrical connection with the power bus wire.

● From the bottom side of the body segment, push the power bus receptacle into the body segment.
It should be flush with the bottom of the body segment.

§§§ If the power bus receptacle is loose, remove it and crimp it gently midway along the section that fits into the body segment.

Make Six Knee Clamp Assemblies

✓ Grasp the knee clamp holder with the needlenose pliers so that the drill bit does not slip.

§§§ Use the drill to ream out the hole that runs through the center of the knee clamp holder.

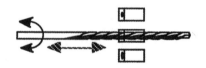

✓ Ream the entire length of the hole.

§§§ Bend the nitinol actuator in half but...

✗ ...do **not** crimp it at the bend.

◉ Insert the free ends of the actuator into the knee clamp holder and push the knee clamp holder toward the bend...

§§§ ...until the bend protrudes about 2 to 3 millimeters.

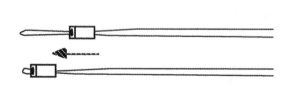

✓ Swage the actuator clamp.

● Press the actuator clamp into the knee clamp holder at the hole where the bend protrudes.

✗ Do **not** use excessive force to press-fit the actuator clamp, or the nitinol will break now, or after a few activation cycles. If it is difficult to press-fit the actuator clamp into the knee clamp holder, remove the actuator, ream out the knee clamp holder again and swage the ends of the actuator clamp.

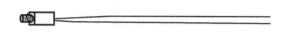

✓ When it is finished, a small loop of nitinol will protrude from the knee clamp holder. The actuator clamp will be flush with the other end of the knee clamp holder.

✓ Note that the nitinol is held between the knee clamp and the knee clamp holder.

✗ The nitinol does **not** run through the knee clamp holder.

Attach Two Horizontal-Vertical Actuators to Each Body Segment

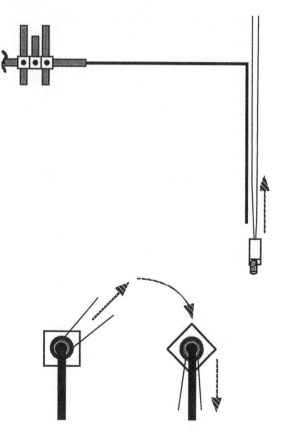

◉ Slide the knee clamp assembly onto the end of the leg. The free ends of the nitinol actuator must exit the knee clamp assembly facing toward the **top** of the body segment.

✓ Push knee clamp all the way to the knee.

✓ The knee clamp should rotate freely on the leg.

✓ Pull outward on the actuator wires, then, while pulling outward, change the direction of pull so that the actuators pull towards the body segment. The knee clamp holder may not be square with respect to the body, but that is OK.

✗ The actuator wires must **not** be crossed.

✗ The actuator wires must **not** cross over the leg, they should instead be free on either side of the leg.

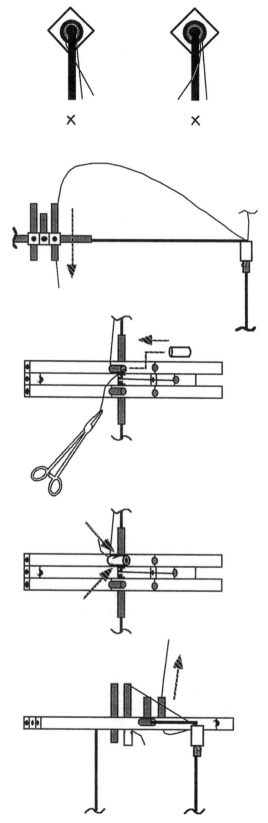

● Now thread the free end of the front-most nitinol actuator through the top of the vertical clamp and out the bottom.

✓ Grip the free end of the actuator with the locking forceps and let it hang to provide tension to the actuator.

● Press-fit the aluminum outer clamp over the nitinol and the actuator clamp. It will be a snug fit.

✓ The nitinol is clamped on the **bottom** of the robot.

✗ Do **not** use excessive force to press-fit the outer clamp. If it is difficult to press-fit the outer clamp, ream out the inside ends slightly.

ʃʃʃ Use the wire cutters to **gently** crimp the outer clamp about halfway down its length so that the nitinol actuator is held tightly.

● Now thread the free end of the other nitinol actuator through the **bottom** of the horizontal clamp and out the top.

✓ Grip the free end of the actuator with the locking forceps and let it hang freely to provide tension to the actuator.

● Press-fit the aluminum outer clamp over the nitinol and the actuator clamp. It will be a snug fit.

✓ The nitinol is clamped on the **top** of the robot.

✗ Do **not** use excessive force to press-fit the outer clamp, or the nitinol will break. If it is difficult to press-fit the outer clamp, ream out the inside ends slightly with the knife.

✄✄✄ Use the wire cutters to **gently** crimp the outer clamp about halfway down its length so that the nitinol actuator is held tightly.

✓ Repeat for the actuator on the other side of the body segment.

✓ A single finished horizontal-vertical actuator looks like this from the side.

✓ Now attach the actuators to each of the remaining body segments.

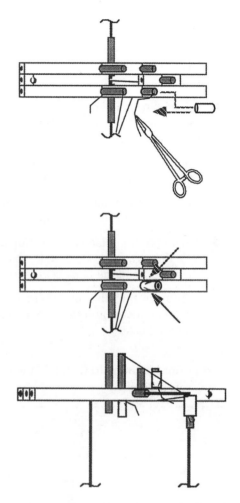

A body segment with the actuators and the power bus attached looks like this (reduced size illustration).

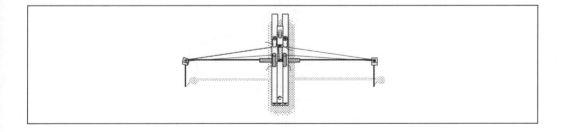

 You can **test the actuators** on each body segment by applying power across the power bus and each actuator clamp. Nitinol is not directional, so it does not matter at this time if the power bus is positive or negative (later we will make the power bus positive because the driver circuit on the interface card pulls the actuator signal to ground).

Pin the Three Body Segments and Two C-Joints Together

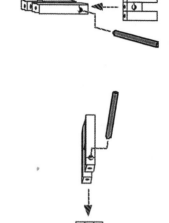

✓ Insert a c-joint to the rear of a body segment...

◉ ...then pin it in place by inserting an axle through the horizontal axle holes in the body segment and the c-joint.

✓ Repeat for the other c-joint and one of the remaining two body segments.

✓ Insert the front end of a body segment into the open end of a c-joint...

◉ ...then pin it in place by inserting an axle through the vertical axle holes in the c-joint and the body segment.

✓ Repeat for the other body segment + c-joint assembly and the remaining body segment.

Your finished Stiquito II should look like this (reduced size illustration).

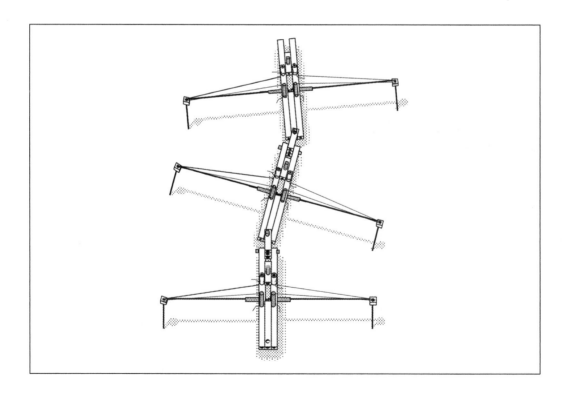

 Congratulations! Your Stiquito II is complete! Now is a good time to **take a break**. Build the IBM PC and compatible computer parallel port interface later.

A Useful Variation: The Rigid Stiquito II

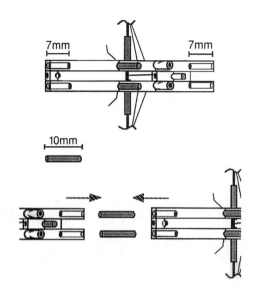

§§§ Use the drill bit to ream out each end of the hole that runs through the center of each body segment side to a depth of about 7 millimeters.

✓ Repeat for each body segment.

§§§ Cut four 10-millimeter body segment pins out of 1/16" outside diameter copper tube.

● Pin the body segments together with body segment pins, using two at each joint between body segments.

✗ The c-joints are **not** used in this version.

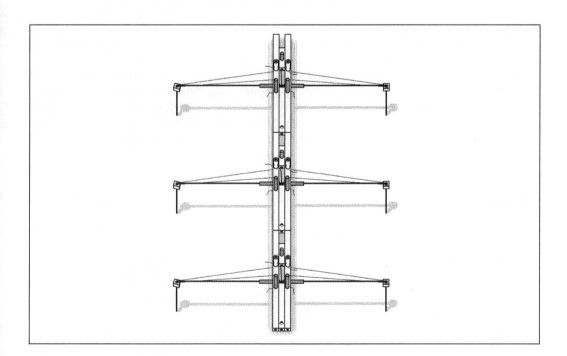

Your finished **rigid** Stiquito II should look like this (reduced size illustration).

The rigid version can carry a miniboard attached underneath the body. Use a c-joint to connect two rigid Stiquito IIs. One can carry a miniboard, the other can carry rechargeable nickel-cadmium batteries to power the miniboard and the robot. Attach a "solar cell sail" to the top of each segment to recharge the batteries and you are ready to send the **Stiquito II Rover** to Mars!

Well, maybe NASA would prefer that you try it out on a gravel driveway first.

Chapter 4

Increasing Stiquito's Loading Capacity

John K. Estell, Thomas A. Owen, and Craig A. Szczublewski

The Stiquito robot possesses a limited payload capacity. This becomes a serious design constraint when attempting to develop an on-board microcontroller system for Stiquito. In order to provide greater stability and to increase the load it can carry, modifications must be made to the Stiquito design. By splitting the hull into two halves, greater stability is achieved. By redesigning the leg system, more weight can be supported. The combination of these two features makes it possible to design a modified Stiquito body that is capable of handling the load requirements of a self-contained microcontroller system.

INTRODUCTION

The Stiquito robot body has a limited payload capability of about 50 grams.[1] In attempting to design a microcontroller board for Stiquito, problems were encountered with balance, weight limitations, and size. In order to add a self-contained controller board, like the SCORPIO system designed at the University of Toledo,[2] alternative body designs must be considered. The research into alternative designs focuses mainly on strengthening the body and legs to increase payload capacity without impeding the nitinol actuators.

The original Stiquito design[3] consisted of a single-hull chassis constructed of two 60-millimeter lengths of $1/8$-inch (3-millimeter) styrene rods. The two lengths were attached side to side to form a chassis 60 millimeters long and 6 millimeters wide. The styrene rod was an excellent choice as it is fairly durable, easy to work with, extremely lightweight, and readily available at low cost. The single-hull design provided a stable base to which everything else was attached. Additional support was offered from each leg pair being formed of a single length of wire, penetrating both styrene rods.

The leg pairs were constructed of 100-millimeter lengths of 0.020-inch (0.5-millimeter) piano wire. The effective length of an individual leg was 30 millimeters long and 10 millimeters high. The leg pairs were placed 23 millimeters apart on the chassis to distribute the payload evenly. The piano wire proved to have considerable strength without overwhelming the nitinol actuators. The electrical connections consisted of a ground connection tied into all the legs via a conductive backbone. The circuit was completed with the attachment of the nitinol actuators to body crimps made from aluminum tubing.

CONSTRAINTS AND MODIFICATIONS

Although the 0.020-inch piano wire provided a stable leg system for the original robot, the SCORPIO project required that the leg system be able to support more than 50 grams. If a 9-volt battery were used as the power supply for the robot, the power supply would require 36 grams[4] of Stiquito's 50-gram loading capacity. A simple solution to the problem would be to increase the number of legs on the robot. This, however, increases the current demand on the power supply. Each leg only adds about 8 grams to the carrying capacity of the robot, but draws an additional 180 milliamps of current.[5] It was clear that a new leg design must be created that would increase the weight-to-current ratio that each leg provided.

The principal idea of the SCORPIO project was to mount on the robot a microcontroller system contained on a perforated circuit board for independent operation. Accordingly, the chassis must provide a stable platform for the board. The single-hull design of Stiquito requires that the payload be balanced within a few grams for reliable operation. Since Stiquito is approximately the size of a 40-pin dual in-line pin (DIP) component, it was obvious that this was not a platform to support a board consisting of several integrated circuit chips and the associated power supply.

The new design overcomes the weight constraints with a redesign of the original leg system. The design features reinforced legs that are capable of carrying a much greater load than the original robot. The legs are mounted on a pin joint that is supported both top and bottom instead of being an integral part of the body. This provides for both freedom of movement in the horizontal plane and reinforcement in the vertical plane.

Improvements to the chassis consist of a twin-hull design that can be spaced at a variable distance. This provides maximum stability for a circuit board and its power supply without compromising either payload capacity or performance. The styrene rods were retained for the chassis construction because of their favorable properties. The length was extended to accommodate the length of the circuit board. With the modified leg and chassis design, it was also necessary to provide a recoil mechanism to return the leg to its rest position. A simple way to accomplish this is to use a rubber band; this will not offer too much resistance to the operation of the nitinol.

CONSTRUCTION

The first step in the construction of the modified Stiquito is gluing the hulls. Each hull of the design is made of five 14-centimeter-long sections of ⅛-inch (3-millimeter) styrene rod. These sections are glued together as shown in Figures 4.1 and 4.2, and allowed to dry overnight. It is important to allow sufficient drying time before any additional work is performed, because much greater stresses are applied to this hull than the original Stiquito hull. If the glue is not allowed to cure fully, the bond will likely fail. The cross-members can be installed after the hull assembly has been completed; they can be of any convenient length, or they can be omitted altogether. The dimensions, given in Figure 4.1, are such that a 9-volt battery and two AA batteries can be supported between the hulls. The 9-volt battery is used in the SCORPIO design to power the control circuitry, and the two AA batteries are used to provide power to the legs.[6]

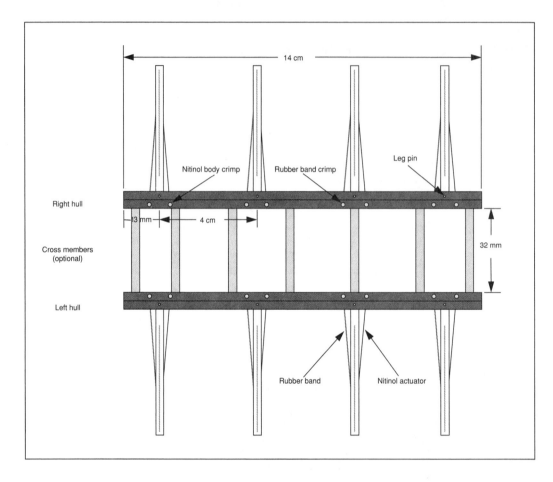

Figure 4.1. Top view of modified Stiquito body.

The legs can be constructed while the cross-members are drying. Each leg is formed from two lengths of styrene rod and an 80-millimeter length of 0.036-inch (0.9-millimeter) piano wire, as shown in Figure 4.2. The construction of each leg is such that the styrene rod provides all of the horizontal strength, while the piano wire forms the knee and lower leg. The longer styrene rod is used as the upper leg. A notch 1.5 millimeters long, cut from the bottom to the hollow center, is made at one end of this rod. This is where the knee is formed by the piano wire, which runs through the center of the rod. By using a notch so that the vertical portion of the wire is still underneath the horizontal rod, greater support is provided to the overall design. Finally, the shorter rod is glued to the horizontal rod to hold the piano wire firmly in place.

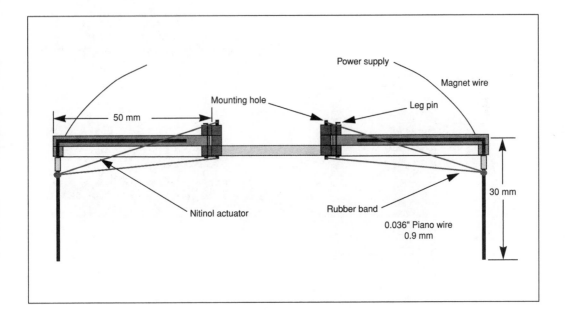

Figure 4.2. Front view of modified Stiquito body.

After the cross-members have fully cured, holes can be drilled in the body. Three holes are required for each leg: one for the nitinol body crimp, one for the rubber band crimp, and one for the leg pin. When drilling the leg pin holes, it is best to have the legs already in position; it is thus possible to ensure that the pin holes are straight from top to bottom, as shown in Figure 4.3. It is also important to make sure that drilling is performed perpendicular to the body so that the hinge is in the proper orientation. If the pin hole is not drilled properly, the leg will bind against the body during use. The nitinol body crimp holes are similar in design and use to the holes in the original Stiquito. The rubber band holes are an addition to the original Stiquito design; rubber bands are used in place of piano wire to provide the recoil force for the nitinol wire. There are no mounting holes for the power wires, as this system was designed to be controlled by an on-board circuit board. The magnet wire that carries the power to the nitinol goes directly from the legs to the circuit board.

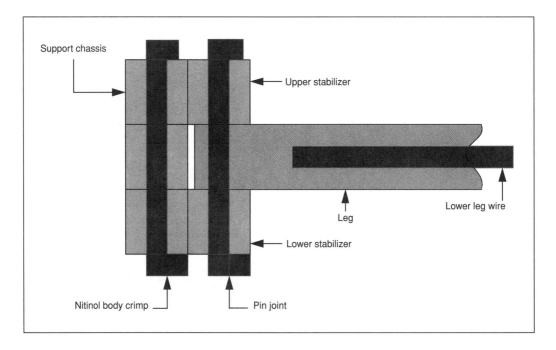

Figure 4.3. Detailed view of modified Stiquito pin joint.

After the holes are drilled, the legs are mounted on the body with short lengths of piano wire used as leg pins. The nitinol actuators, power wires, and crimps are now installed. (It is not necessary to actually crimp the leg crimps, as the nitinol and magnet wire will not slip between the leg and the crimp.) When this is completed, the rubber bands are installed by knotting a free end of each rubber band around its respective leg, passing the rubber band through the mounting hole, and knotting the other end around the leg again. All that remains to be done is bending the ratchet feet. This completes the construction of the modified Stiquito body.

Sources of SCORPIO building materials can be found in Appendix C.

CONCLUSION

The split-hull system provides both a stable platform for a circuit board and a convenient location for the power supply. Because of this and the redesigned legs, the modified Stiquito body has held up well in testing. It has successfully supported a load of 400 grams on eight legs. This compares favorably to the approximately 70 grams that an eight-legged version of the original Stiquito design could support. The modified body has a top speed of approximately 6.7 centimeters per minute carrying a 300-gram payload. As the SCORPIO microcontroller system weighs considerably less than this, the new body design can support enough weight to allow for future expansion. With the combination of increased payload and improved balance, the modified Stiquito body has proven itself to be a suitable base for robotics experimentation while still being affordable for hobbyists and laboratory experimenters.

REFERENCES

1. See Chapter 2.

2. See Chapter 10.

3. Mills, J.W. 1992. Stiquito: A small, simple, inexpensive hexapod robot. Computer Science Dept., Indiana Univ., Technical Report No. 363a.

4. Radio Shack. 1990. *Enercell battery guidebook*, 2d ed. Richardson, Tex: Master Publishing, Inc.

5. Gilbertson, R. 1992. *Motorless motion! Working with shape memory wires*, 2d ed. San Anselmo, Calif.: Mondo-tronics, Inc.

6. See Chapter 10.

Chapter 5

Boris: A Motorless Six-Legged Walking Machine

Roger G. Gilbertson

Figure 5.1. *Boris*, A Motorless Six-Legged Walker. Following in the footsteps of larger and more sophisticated walking machines (like *Genghis* and *Attila* from MIT), *Boris* presents an exciting, fully controllable and low cost project that can be up and "running" quickly. And because Boris uses motorless Muscle Wires® technology it has great potential for expansion and adaptation in many ways.

Excerpted from the "Muscle Wires Project Book" by Roger G. Gilbertson ©1994 Mondo-tronics, Inc. ISBN 1-879896-13-3. Muscle Wires is a registered trademark of Mondo-tronics.

After you understand and have practiced the basics of how to connect power and utilize the strength of Muscle Wires (nitinol) as shown in the previous projects, you're ready to build Boris, our tiny, motorless walking machine.

Boris moves with an amazingly smooth and lifelike action, taking steps about one centimeter long and one centimeter high.

Boris's simple design uses eight separate Muscle Wires to create all motions; one to make the forward/backward motion on each of the legs, and two wires for lifting alternating legs with each step. Build Boris from piano wire, balsa wood or foam core, and plastic sheet. Connect Boris via a lightweight wire wrap wire harness to an interface circuit that accepts signals from the parallel printer port of a PC-type computer. Then operate Boris from the computer's keyboard.

Boris Parts List	
Amount	**Item**
70 cm	Piano wire, 0.38 mm (0.015″) dia.
14 cm	Piano wire, 1.25 mm (0.050″) dia.
16 each	Hex nuts, 4-40
16 each	Screws, 4-40, ⅛″
12 cm	Plastic tube, 3 mm (0.125″) OD, 1.5 mm (0.062″) ID
36 cm	Plastic tube, 2.5 mm (0.10″) OD, 1.25 mm (0.050″) ID
60 cm	Flexinol Muscle Wire, 100 μm (0.004″) dia. Cut as needed
1 each	Header, 8 pin, 2.5 mm (0.100″) spacing
	Balsa wood or foam core, 6 mm (0.25″) thick
	Plastic sheet, 1.25 mm (0.050″) thick
	Wire wrap wire
	Silver solder and Flux
	Epoxy, 5 minute type

ASSEMBLE BORIS

Cut and bend the thin piano wire (a very springy steel wire available at hobby stores) into the rectangle shown in the Pattern Sheet (page 120). Solder the rectangle closed using silver solder and its flux (available at hobby and craft stores).

The flux cleans the wire for a solid bond. Follow its instructions and observe all safety precautions. When soldering, tape the parts down to hold them. In the same joint solder a 12 cm length of insulated wire wrap wire as shown. Repeat for six rectangles.

Solder a hex nut to a wire rectangle (Figure 5.2), placing it as shown. Hold the pieces in place with tape on a cardboard or wood base, and use silver solder and flux. Make three left and three right legs. Be sure to not let any solder run inside the threads of the nut.

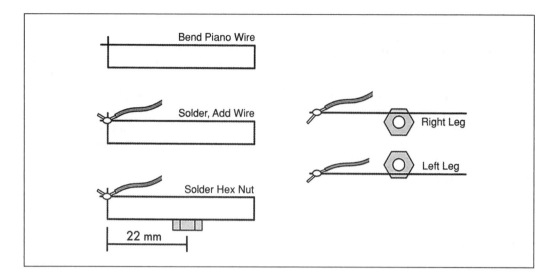

Figure 5.2. Leg Components. Thin piano wire provides both structure and a bias force for each leg. Note position of nut for left and right side legs.

For the ten remaining hex nuts, solder a 12 cm length wire wrap wire to each, as in Figure 5.3.

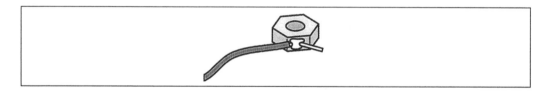

Figure 5.3. Solder Wires. Be sure to not let solder run inside the nut. (If this happens, replace the nut.)

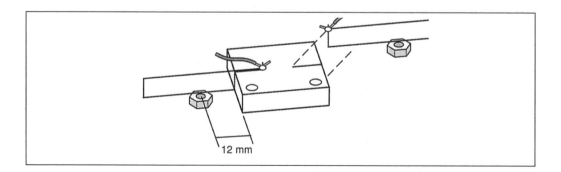

Figure 5.4. Leg Module Assembly. Insert two wire rectangles into the slots in the leg module square. The nuts must be located 12 mm from the edge, as shown. Epoxy the wires in place.

Make the three leg modules from balsa wood or foam core, with slots and holes as shown on the Pattern Sheet (page 119).

Place the left and right wire rectangles into the leg modules, locating the hex nuts as shown (Figure 5.4). The top and bottom of the rectangles should be flush with the surface of the module. Glue them in place with a "5-minute" type epoxy. Repeat to make three identical leg modules.

Prepare the ten hex nuts (the ones with wires soldered to them) by covering the hole on each nut with a narrow strip of tape to prevent epoxy from getting into their threads. Repeat for both sides of all ten nuts.

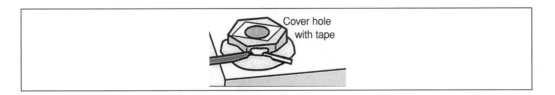

Figure 5.5. Epoxy the Hex Nuts to the Leg Module. Do not get glue inside the nut or the hole below it.

Epoxy a hex nut to the foam core leg module, over a pre-cut hole as in Figure 5.5. Repeat for the second hole on the same side, and all three leg modules.

Use care not to get epoxy in the holes on the leg module or inside the hex nut. If epoxy should get on top of the nut, wipe off the excess, then when fully hardened, sand the top of the nuts with fine sandpaper.

Epoxy another two hex nuts to the foam core, on the opposite side from the other nuts, on just the front and back leg modules (Figure 5.6). Remove the protective tape from all the hex nuts.

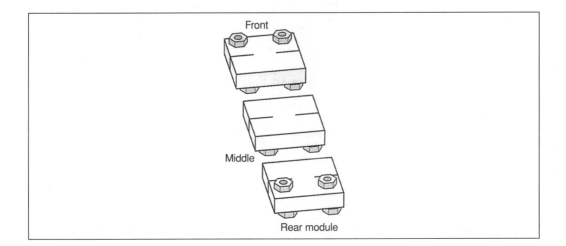

Figure 5.6. Nuts on Top. Locate nuts on the front and rear modules only, and epoxy in place.

Into the hole that runs through the center of each leg module, insert a 4 cm long 3 mm diameter plastic tube, centering it so both ends extend the same amount, and epoxy in place (Figure 5.7). Repeat for all three modules.

Figure 5.7. Insert the Large Plastic Tube. Glue tube in place with epoxy, centering it in the foam core. Repeat for all three leg modules.

Slide the three leg modules onto the 14 cm long piano wire center rod, orienting each as in Figure 5.8. Let the rod extend evenly from each end and epoxy it to the front and back modules. Adjust for a small amount of space (0.5 mm) so the center module can spin freely.

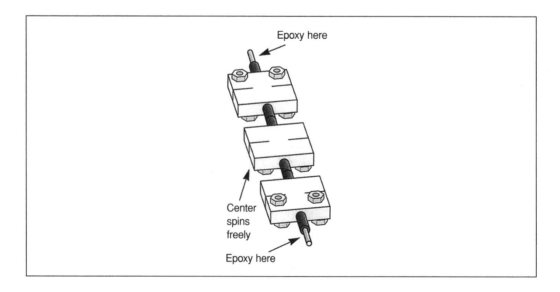

Figure 5.8. Place All Three Leg Modules on Center Rod. With epoxy glue the front and back modules to the rod, spacing them slightly apart so the center module can rotate freely. (Wires omitted for clarity.)

Cut out the leg panels from 1.25 mm sheet plastic and epoxy them to the wire rectangles (Figure 5.9).

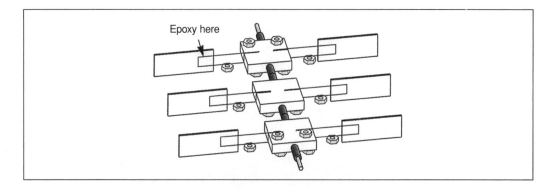

Figure 5.9. Attach Leg Panels with Epoxy. Hold the leg panels in place with tape until the epoxy hardens. Be sure they are all square and even to each other. (Electrical wires omitted for clarity.)

Hold them with tape until the epoxy hardens. Be sure they are parallel and even to each other.

Place the whole assembly on a spacer like the cassette boxes and pencils in Figure 5.10. Tape the 5 cm long leg tubes to the leg panels, as shown. Adjust the tubes up and down so they all contact the table surface with the leg modules resting flat on the pencils. Glue the tubes to the panels with epoxy or a solvent glue.

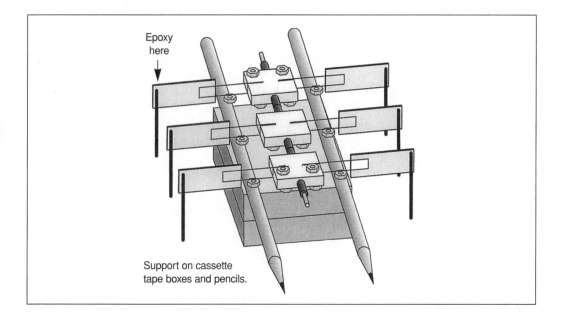

Figure 5.10. Attach Leg Tubes with Epoxy. Place the assembly on a supporting block and adjust so all legs are the same length. Hold tubes with tape until dry.

Form the Spacer Block (as on the Pattern sheet) from balsa wood or foam core, and glue to the middle leg module (Figure 5.11). Epoxy two 1 cm long segments of tube to the top of the block. Let it rest until all glue has hardened.

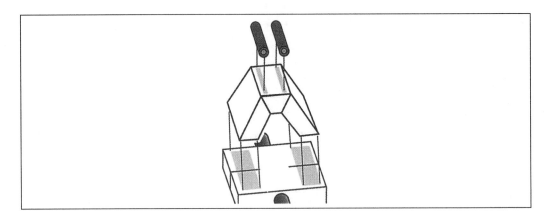

Figure 5.11. Spacer Block. Form the spacer block and epoxy it and two tubes to the middle leg module.

Install the Muscle Wires

Install screws in all sixteen hex nuts. Attach a length of 100 μm diameter Muscle Wire to one of the legs by catching it between the nut and screw.

Tug on the Muscle Wire and position it under the screw on the OPPOSITE side of the module, as in Figure 5.12. Keep enough tension on the wire to bend the leg about five degrees, then tighten the screw to secure the wire. This tension provides the bias force needed to extend the Muscle Wire after it cools. Trim the wire, leaving about 1 cm extra for later adjustment and repeat for all legs.

On each module, note how the wire pairs cross. To prevent them from short circuiting, separate them with a 1 cm by 2 cm slip of paper held with tape, as in Figure 5.12.

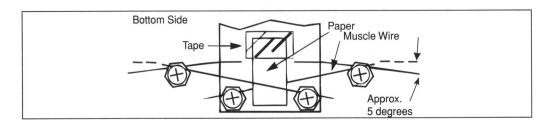

Figure 5.12. Install the Muscle Wires on the Legs. Catch the wire under the screw head, and tense it to bend the leg a bit, as shown. Insulate with paper slip.

On the top side, attach a length of Muscle Wire for the left side and right side lift, as in Figure 5.13. Note that these wires pass through the short tubes on the spacer block, and *do not* cross. Adjust these wires so they have just enough slack to let the middle leg module move up or down about 1 cm, as measured at the feet. Attach the header, as in Figure 5.14.

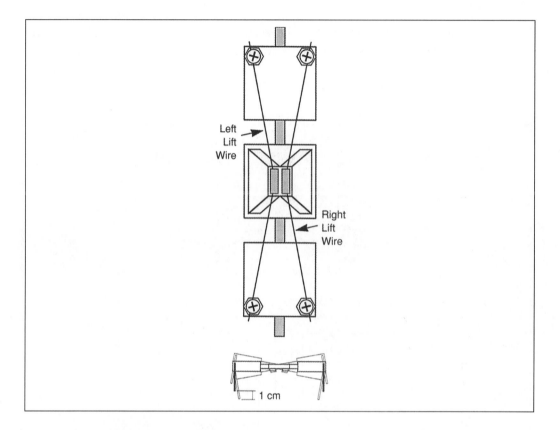

Figure 5.13. Lift Wires. These wires pull the middle leg module from side to side to make the leg lifting action. Leave just enough slack so the legs move up and down about 1 cm.

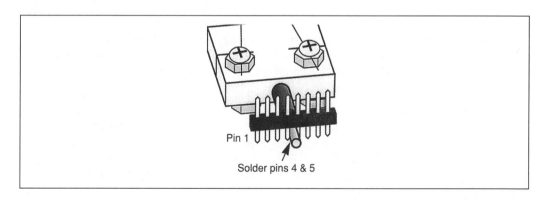

Figure 5.14. Attach Header. Solder pins 4 and 5 to center rod for a good electrical connection and a solid physical one, too.

Electrical Wiring

To prevent confusion and incorrect wiring, review the diagrams, and double check each connection carefully, both before and after making it. Note that the electrical wires connect to the 8-pin header block, permitting easy connection of the robot to the interface circuit described below.

Figure 5.15 shows Boris's physical and electrical connections. Connect the wires as listed in the table. Note that the center rod acts as a common power supply to all the Muscle Wires. Double check all connections.

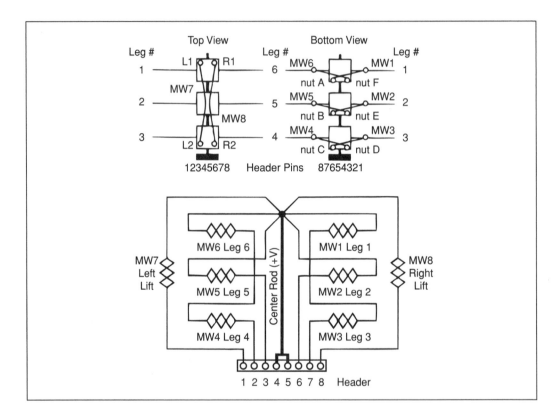

Figure 5.15. Wiring Diagram and Connections. Viewing Boris from above, number the legs from 1 to 6 like an integrated circuit, as shown. The schematic shows the six leg wires, and two lift wires. Header pins 4 and 5 connect to the center rod which powers all Muscle Wires.

Connections on top side:
Hex nut wires L1 and R1 to Rod.
Hex nut wire L2 to Header pin 1.
Hex nut wire R2 to Header pin 8.

Connections on bottom side:
Hex nut wire A to Leg wire 3.
Hex nut wire B to Header pin 6.

Hex nut wire C to Header pin 7.
Hex nut wire D to Header pin 2.
Hex nut wire E to Header pin 3.
Hex nut wire F to Leg wire 4.
Leg wires 1, 2, 5 and 6 to Rod.
Solder Header pins 4 & 5 to Rod.

Interface Circuit

In order to make Boris walk, you will need to build an interface circuit. You can obtain these supplies from an electronics store or from one of the vendors listed in Appendix C.

Parallel Port Interface Circuit Parts List		
Amount	**Part Number**	**Item**
8	R1-8	Resistor, 10KΩ, $^1/_4$ W (Brn, Blk, Org, Gold)
9	R9-17	Resistor, 100Ω, $^1/_4$ W (Brn, Blk, Brn, Gold)
1	R18	Resistor, 1KΩ, $^1/_4$ W (Brn, Blk, Red, Gold)
2	R19,20	Resistor, 5Ω, $^1/_4$ W (Grn, Blk, Gold, Gold)
1	VR1	Variable Resistor, 2KΩ, $^1/_4$ W
8	I1-8	LED, Red, T-1 size
1	I9	LED, Green, T-1 size
8	Q1-8	Transistor, MPS A13 NPN, Darl. Amp. (Mouser #333-MPSA13)
1	U1	IC, 317T Regulator (Radio Shack #276-1778)
1	SW1	Switch, SPST (Single Pole Single Throw)
1	J1	Connector, D-25 male
1	J2	Connector, 8 pin SIP (or half a 16 pin DIP socket)
1	J3	Connector, Power Jack (as needed)
8 meters		Wire Wrap Wire
9 meters		Hookup Wire
		Perf Board (see Figure 5.16)
		Solder

Build the circuit in Figures 5.16 and 5.17. It receives signals from a computer's parallel printer port and provides an adjustable voltage to Boris's Muscle Wires. Build the circuit with the wire wrap or perf board methods.

Build the harness with seven wire wrap wires, each about 1 meter long. Bundle them together with small lengths of wire wrap wire or with loops of thread.

When assembled, test the circuit, carefully measuring and adjusting the output before attaching it to Boris. Start a BASIC program, then enter and run Program Three (Figure 5.18). Leaving switch SW1 *off* until all signals have been checked, test that each bit can be turned on and off from the keyboard.

At this point you may wish to connect Boris and test each action. This provides a good way to check for correct wiring, power levels, and proper leg actions.

Finally, run Program Four (Figure 5.19), connect Boris, and *watch it walk!*

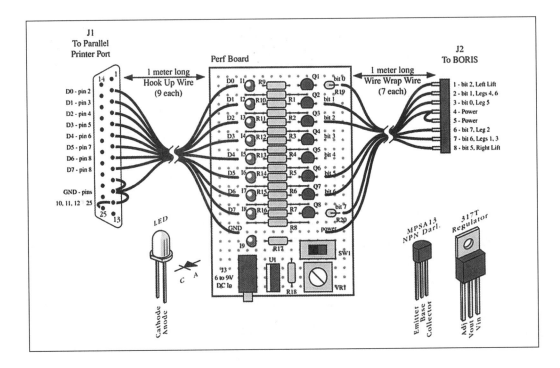

Figure 5.16. Boris Interface Circuit and Cables. Use this component layout to easily connect the parts as shown in Figure 5.17. Connect J1 to the parallel port on a PC-type computer and connect J2 to Boris. Power the circuit with a 6 to 9 V DC supply, and adjust the variable resistor VR1 to set the power level to the Muscle Wires. LEDs I1-8 show the bit pattern on the parallel port.

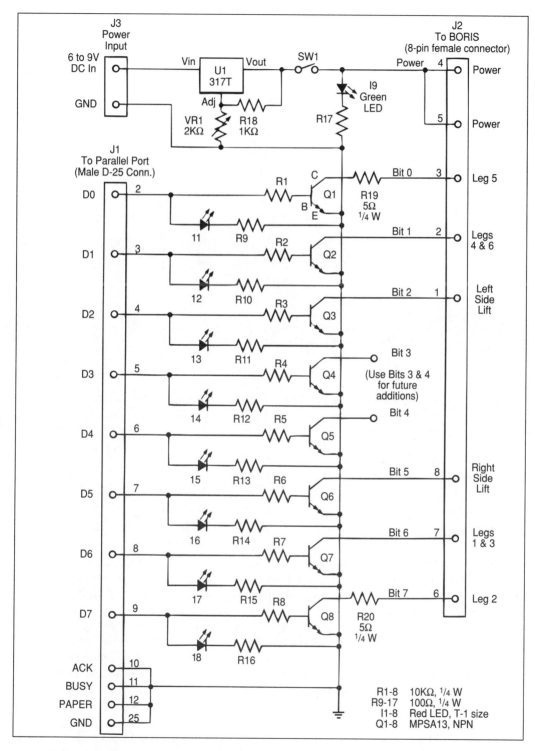

Figure 5.17. Schematic for Parallel Port Interface Circuit. This circuit captures the bit pattern from a standard PC-type parallel port and displays it with LEDs (ON = 1, OFF = 0). Also, each transistor, Q1 to Q8, provides power to drive a Muscle Wire when its corresponding bit is high. Boris uses six of the eight circuits. Output Bits 3 and 4 remain free for future additions (grippers, feelers, pointing mechanisms, etc.).

```
     NEW

     10    REM Test Program Three - Blip A Line
     11    REM 9210.19 Rev 9304.31
     12    REM Activate a line on the parallel port for short time
     13    REM
     14    REM   Keyboard    Output      BORIS
     15    REM   "1" Key  =  Bit 0  =  Leg 5
     16    REM   "2" Key  =  Bit 1  =  Legs 4 & 6
     17    REM   "3" Key  =  Bit 2  =  Left Side Lift
     18    REM   "4" Key  =  Bit 3  =  not used
     19    REM   "5" Key  =  Bit 4  =  not used
     20    REM   "6" Key  =  Bit 5  =  Right Side Lift
     21    REM   "7" Key  =  Bit 6  =  Legs 1 & 3
     22    REM   "8" Key  =  Bit 7  =  Leg 2
     23    REM
     30    DELAY = 200: REM Delay ON time, Increase for longer
     40    LPRINT CHR$(0); : REM Clear port
     50    PRINT "Press 1 thru 8 to activate bits, Q to end"

     100   A$ = INKEY$: REM Get key press
     110   IF A$ = "1" THEN LPRINT CHR$(1); : GOTO 200: REM bit 0
     120   IF A$ = "2" THEN LPRINT CHR$(2); : GOTO 200: REM bit 1
     130   IF A$ = "3" THEN LPRINT CHR$(4); : GOTO 200: REM bit 2
     140   IF A$ = "4" THEN LPRINT CHR$(8); : GOTO 200: REM bit 3
     150   IF A$ = "5" THEN LPRINT CHR$(16); : GOTO 200: REM bit 4
     160   IF A$ = "6" THEN LPRINT CHR$(32); : GOTO 200: REM bit 5
     170   IF A$ = "7" THEN LPRINT CHR$(64); : GOTO 200: REM bit 6
     180   IF A$ = "8" THEN LPRINT CHR$(128); : GOTO 200: REM bit 7
     190   IF A$ = "q" OR A$ = "Q" THEN LPRINT CHR$(0); : END
     195   GOTO 100

     200   REM Delay loop
     210   FOR I = 1 TO DELAY
     220   NEXT I
     230   LPRINT CHR$(0); : REM After delay, clear output
     240   GOTO 100
```

Figure 5.18. Listing for Test Program Three. Enter this listing as a BASIC program and run it on your computer. Use it to test each Muscle Wire individually, and to debug the interface circuit.

How It Works

When the BASIC software running in the computer sends a command to the parallel port, it creates a pattern of eight bits, which the parallel port represents as voltages on eight wires (D0 through D7). A high signal (near +5 Volts) represents a 1, and a low voltage (near 0 Volts) represents a zero.

The BASIC command "LPRINT CHR$(bits);" sends a single character of eight ones and zeros. The actual pattern depends on the number entered for "bits," from 0 to 255. The semicolon (;) at the end of the command line holds, or "latches," the bit pattern on the port. (If the LPRINT command line did not end in a semicolon, the program would send a Line Feed and a Carriage Return character, which would confuse the robot's walking pattern.)

The eight bits connect to the interface through J1, the D-25 connector, and eight hook up wires. A ninth wire provides a common ground between the computer and board. Each of the eight bits functions in the same way. For example, when Bit D0 goes high, it causes LED I1 to glow, and turns on transistor Q1 which provides power, via the wire wrap wire harness, to Leg 5 of Boris.

Resistors R19 and R20 equalize the total resistance of the circuits for Legs 2 and 5. Without them, the lower resistance would prevent full contraction of other Muscle Wires

```
NEW
10    REM Test Program Four - RG 9305.11
11    REM Walking Program for BORIS Connected to PC Parallel Port
12    REM Control from numeric keypad: UP arrow = Forward, etc.
13    REM
14    REM    Send bits to the parallel port with    LPRINT CHR$(Bits);
15    REM
16    REM    Bits = 1*n + 2*n + 4*n + 8*n + 16*n + 32*n + 64*n + 128*n
17    REM
18    REM Bits equation sets bits ON if n = 1 or OFF if n = 0
19    REM    First position (1*n) is bit 0
20    REM    Last position (128*n) is bit 7
21    REM Calculate with each n to 1 or 0 for other output patterns
22    REM
23    REM Note: Ending semicolon (;) on LPRINT command holds
24    REM         bit pattern at parallel port until next LPRINT
25    REM
30    DELAY = 350: REM Increase for slower walk, decrease for faster
40    PRINT "Press keypad arrow keys for Fwd, Back, Left, Right"
50    PRINT "Press space bar for Stop"
60    PRINT "Press Q at any time to Quit"

100   REM Stop walking
110   LPRINT CHR$(0);      : GOSUB 1000: REM Clear All Bits
120   GOTO 100

200   REM Walk Forward
210   LPRINT CHR$(65);   : GOSUB 1000: REM Bit 6, 0 ON
220   LPRINT CHR$(4);    : GOSUB 1000: REM Bit 2 ON
230   LPRINT CHR$(130);  : GOSUB 1000: REM Bit 7, 1 ON
240   LPRINT CHR$(32);   : GOSUB 1000: REM Bit 5 ON
250   GOTO 200

300   REM Walk Reverse
310   LPRINT CHR$(65);   : GOSUB 1000: REM Bit 6, 0 ON
320   LPRINT CHR$(32);   : GOSUB 1000: REM Bit 5 ON
330   LPRINT CHR$(130);  : GOSUB 1000: REM Bit 7, 1 ON
340   LPRINT CHR$(4);    : GOSUB 1000: REM Bit 2 ON
350   GOTO 300

400   REM Rotate Left
410   LPRINT CHR$(66);   : GOSUB 1000: REM Bit 6, 1 ON
420   LPRINT CHR$(32);   : GOSUB 1000: REM Bit 5 ON
430   LPRINT CHR$(129);  : GOSUB 1000: REM Bit 7, 0 ON
440   LPRINT CHR$(4);    : GOSUB 1000: REM Bit 2 ON
450   GOTO 400

500   REM Rotate Right
510   LPRINT CHR$(66);   : GOSUB 1000: REM Bit 6, 1 ON
520   LPRINT CHR$(4);    : GOSUB 1000: REM Bit 2 ON
530   LPRINT CHR$(129);  : GOSUB 1000: REM Bit 7, 0 ON
540   LPRINT CHR$(32);   : GOSUB 1000: REM Bit 5 ON
550   GOTO 500

1000  REM Check Key Press and Do Delay Loop
1010  A$ = INKEY$
1020  IF A$ = " " THEN PRINT "Stop": LPRINT CHR$(0); : GOTO 100
1030  IF A$ = CHR$(0) + "H" THEN PRINT "Forward": LPRINT CHR$(0); : GOTO 200
1050  IF A$ = CHR$(0) + "P" THEN PRINT "Reverse": LPRINT CHR$(0); : GOTO 300
1060  IF A$ = CHR$(0) + "K" THEN PRINT "Left"    : LPRINT CHR$(0); : GOTO 400
1070  IF A$ = CHR$(0) + "M" THEN PRINT "Right"   : LPRINT CHR$(0); : GOTO 500
1080  IF A$ = "Q" OR A$ = "q" THEN LPRINT CHR$(0); : END

1100  FOR I = 1 TO DELAY
1110  NEXT I
1120  RETURN

1900  END
RUN
```

> Text listings of both of these programs are available free by sending an email request to
> *programs@mondo.com*

Figure 5.19. Listings for Test Program Four. This program permits the operator to control Boris with simple keyboard commands. Pressing the Up arrow on the numeric keypad (#8) causes Boris to walk. The Down arrow makes it walk in reverse, and the Left and Right arrows cause it to rotate. Options include adding gripper hardware to Bits 3 and 4, and expanding these instructions to make a system that can walk around, pick up objects, and perform other telerobotic tasks.

when activating two or more circuits together. Resistors R1 to R8 limit the base current of the transistors, and R9 to R16 limit the current through LEDs I1 to I8.

The 317T regulator, U1, together with variable resistor VR1 and resistor R18, controls the voltage from the power supply connected at J3. Adjust VR1 to increase or decrease the voltage, and thus the power, to Boris.

Caution: Unwanted bit patterns may appear on the parallel port when first turning on the computer or when running other programs. Use SW1 as a "safety switch" to turn Boris's power OFF when not in use. With switch SW1 ON, LED I9 lights, and power flows to Boris as directed by the incoming bit pattern.

As the program runs, it changes the bit patterns at the parallel port, activating different transistors, which in turn activate the various Muscle Wires and make Boris walk.

With properly timed and sequenced patterns of bits (like Test Program Four), Boris can walk forward, go in reverse, and turn in either direction at your command!

Figure 5.20. Boris. Just the first step.

VARIATIONS

The possible variations for Boris are limitless:

- Add foot pads for different terrains; rocky, powdery or carpeted surfaces need larger diameter pads, smooth surfaces need smaller ones; use non-slip rubber for very smooth surfaces.

- Create a lightweight shell and leg covers with an exotic design.

- Add a manipulator and control it via Bits 3 and 4 to pick up ping-pong balls.

- Add sensors on the footpads and legs to make a true sensory/actuator robot system that can avoid objects.

- Add tiny batteries and an on-board controller for an autonomous roving robot not limited by the harness length.

- Make a waterproof Boris that can swim or walk along underwater. (Hint: faster cooling rates mean faster actions!)

- Add a tiny camera and broadcast signals back from remote locations.

- Add two cameras and send back 3-D images.

- Make the cameras waterproof, add lights and send it down the deep water wells beneath Texas to explore the sunless ecology recently found there.[1] Send back pictures and videos of the mysterious inhabitants, and investigate their source of energy.

- Build a team of remote controlled, solar-powered Boris robots and send them to the Moon with a low-cost *Pegasus* launcher via an efficient *fuzzy boundary* transfer orbit.[2] Sell time to adventurers on Earth who can then explore the Moon from their living rooms!

- Send another team of robots with more intelligence outward to explore the deserts, canyons and ice caps of Mars.

- Then prepare for the dozens of amazing worlds circling the giant planets. The stars await!

Like all great journeys, this too, begins with but a single step. Footpads ready...
Go!

REFERENCES

1. E. Pennisi, "Saving Hades' Creatures." *Science News*. Vol. 143, p. 172, Mar. 13, 1993.

2. E. Belbruno, "Through the Fuzzy Boundary: A New Route to the Moon." *Planetary Report*. Vol XII, No. 3, May/June 1992, p. 8. The Planetary Society, 65 North Catalina Ave., Pasadena, CA 91106 USA.

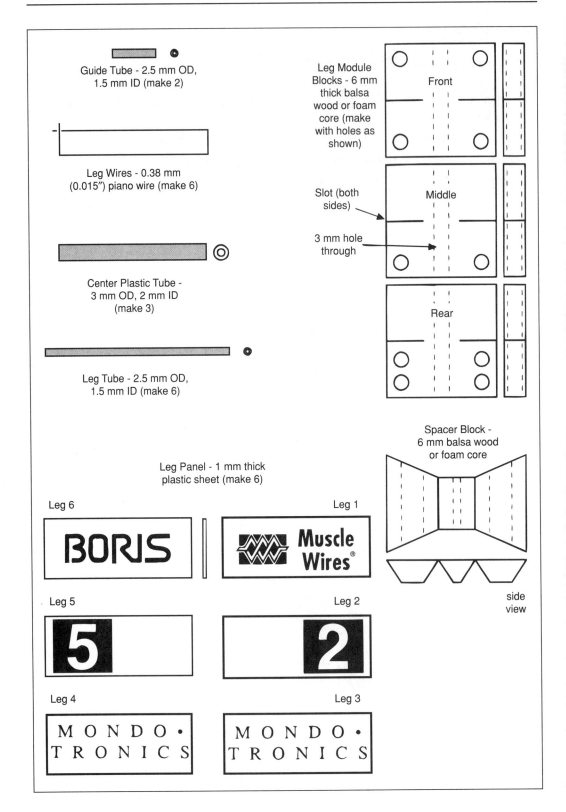

Pattern sheet for Boris, the Motorless Walking Machine.

Chapter 6

A PC-Based Controller for Stiquito Robots

Jonathan W. Mills

PARALLEL INTERFACE FOR IBM PC AND COMPATIBLE COMPUTERS

Stiquito II and Tensipede must be controlled by a computer or other electronic or mechanical device. These robots have 10 or 12 actuators, too many to control manually. This interface allows you to use an IBM PC or compatible computer to control the actuators on Stiquito II or Tensipede and experiment with various gaits. The interface can also be modified to allow sensor inputs to be fed back to the computer, although this is not described in this technical report.

The interface is simple, uses inexpensive parts, is optimized to produce a tripod gait in Stiquito II, and can be fabricated on a single-sided, robust printed circuit board. The interface is attached to the parallel printer port on an IBM PC or compatible computer, one of the most readily available computers in the world. The control programs listed in this report are written in QBASIC, a dialect of BASIC provided free of charge by Microsoft with most IBM PC and compatible computers. The programs are also compatible with earlier versions of BASIC, such as GWBASIC.

The interface card and the BASIC programs described in this report reflect a reasoned and deliberate attempt to provide a "lowest common denominator" for legged robot control. More sophisticated approaches, such as the analog VLSI Łukasiewicz logic arrays and Kirchhoff machines studied by the author, can be found in the literature. Remember that Stiquito II and Tensipede are platforms for research. To derive the most benefit from these simple robots you should use them as a foundation to develop more sophisticated robots and controllers.

Tools and Rules

Additional tools beyond those needed for Stiquito II are needed to build the interface because you must solder the integrated circuits and the control wires to the card. You may want to provide sockets for the integrated circuits. Sockets, used primarily to allow an integrated circuit to be replaced easily, should not be needed if you take reasonable care building and operating the interface card.

More Tools—The following tools are **required** to build the interface card.

- Needlenose pliers
- Wire cutters
- Small hobby knife (X-Acto type)
- Soldering iron
- Solder
- Volt-ohmmeter
- 320-grit fine sandpaper

More Ground Rules

     Look for these cues (small pictures) as you follow the instructions to assemble the interface card. Fewer cues are needed because you are not fabricating parts to close tolerances. A new cue, the soldering iron, has been added to indicate that soldering is involved in the assembly step.

Cue   Meaning

✓   **Do** take the action indicated. No particular precision is required.

✗   **Don't** do this! It is a mistake.

🖎   **Solder the part** at the point or points indicated.

$$$ **Some variation is allowed.** The dimensions are not critical. Stay within one or two millimeters of the specified measurement.

◉   **A loose or sloppy fit is OK.** Your interface will work fine if the fit is not tight.

Precautions

✓   Soldering irons get **very hot.**

✗   Do **not** touch exposed metal on a hot soldering iron. Hold it by the insulated handle.

✗   Do **not** lay a hot soldering iron on the work surface or flammable material.

✓   Use a soldering iron holder or lay a piece of wood under-but-not-touching the tip to protect your work surface.

✓   Although the integrated circuits used in the interface are not particularly static sensitive...

✗   ...do **not** handle them by the pins.

✓   In winter and dry weather, touch a large metal object (table, door frame) to discharge any static electricity before handling integrated circuits.

Now let's build the interface card and use it to control your robot.

Identify Parts

Make sure you have all the parts. Match them against the illustration to be sure they are the correct parts. The printing on the integrated circuits may not exactly match the illustration, but the part number (or 10 KΩ for the resistor package) should appear somewhere on the integrated circuits.

Quantity	Part number; name	Illustration
2 ea	DS2001 Darlington high-current driver	
2 ea	74LS373N Octal tri-state D flip-flop	
1 ea	G5102 5-element 10KΩ resistor network	
1 ea	225M-ND DB-25 male connector	
24 in	HC09G-100-ND 10-wire gray ribbon cable	
1 ea	n/a Printed circuit board (wiring side)	

You will also need hook-up wire and jumper wire. Wire is readily available at electronics and hobby shops, such as Radio Shack.

Solder the Integrated Circuits to the Interface Card

Soldering—Soldering connects integrated circuits to the interface card mechanically and electrically. The basic technique for soldering is described below. You may want to practice on an old circuit board before you solder the parts to the interface card.

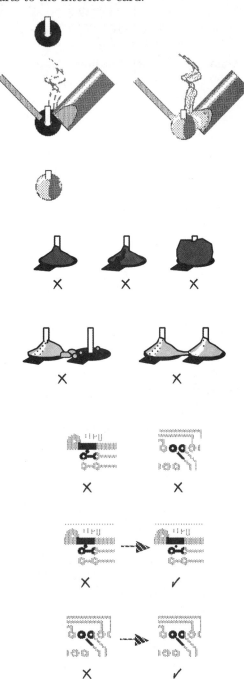

✓ Use the soldering iron's tip to heat the **pad**, not the integrated circuit pin. When the pad is hot, touch the solder to the heated pad, and the solder will flow onto the pad and the pin.

✓ Use just enough solder to wet the pin and cover the pad.

✓ Each solder joint should be bright, shiny, and have flowed evenly around the pin on the pad. The solder on adjacent pads must not touch.

✗ A solder joint should **not** be dull, cracked, or beaded up on the pad.

✗ A solder joint must **not** cross between two pads, or a pad and a trace. This will create a short circuit. Your interface card will almost certainly **not work correctly.**

Check the Board for Short Circuits and Broken Traces

✓ Examine the wiring side of the interface board. Look at places where one trace or pad is near another; check that they do not touch. Look at long traces and near bends; check that the trace is not broken at that point.

✓ If traces or pads touch, but they should not, use the knife to cut the unwanted connection.

✍ If a trace is broken, lightly sand it on either side of the trace, then solder the broken ends together using a piece of fine wire to bridge the gap.

Solder Integrated Circuits to the Board

 Avoid mistakes by fitting all integrated circuits to the board to make sure they are correctly placed before you solder them to the board.

✓ Locate pin 1 on each part. It is the top left pin on each integrated circuit. The top of the integrated circuit is marked by a light-colored band, a U-shaped depression at one end of the integrated circuit, and/or a small circular depression next to pin 1 on the body of the integrated circuit.

✓ An arrow points to pin 1 in the illustration.

◉ Insert the DS2001 integrated circuits into the holes on the component side of the interface board.

✓ There is no wiring on the component side of the board.

✓ The holes into which the integrated circuit pins are inserted are black in the illustration.

✓ The outline of the DS2001 integrated circuits shows their orientation, and arrows in the figure point to the pin 1 locations.

✓ Pins 2, 3, 18, and 19 on each 74LS373 are not used. Bend them upward and outward so that the chip can be inserted into the interface board.

◉ Insert the 74LS373 integrated circuits into the holes on the component side of the interface board.

✓ The holes into which the integrated circuit pins are inserted are black in the illustration.

✓ The outline of the 74LS373 integrated circuits shows their orientation, and arrows in the figure point to the pin 1 locations.

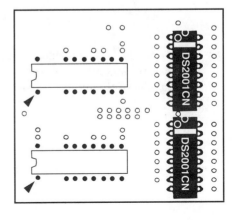

◉ Insert the 10KΩ resistor package into the holes on the component side of the interface board.

✓ The holes into which the resistor package pins are inserted are black in the illustration.

✓ The arrow in the figure points to the location of pin 1 and shows the orientation of the resistor package.

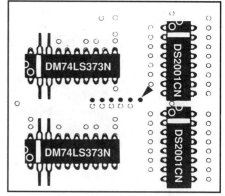

✓ The component side of the interface card should look like this after the integrated circuits and the resistor package are inserted.

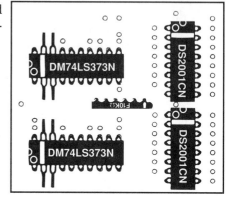

✓ Next, **carefully** turn the interface card over, keeping the integrated circuits pressed firmly to the surface of the card, so that the wiring side is uppermost.

✓ To hold the integrated circuits in place, bend pins 1 and 11 on each 74LS373, pins 1 and 9 on each DS2001, and pins 1 and 6 on the resistor package toward the surface of the card.

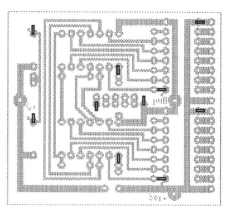

🖎 Solder all the pins on each integrated circuit and the resistor package to the pads on the wiring side of the interface card.

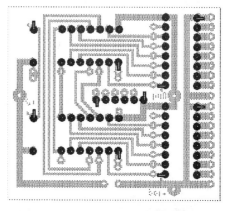

Attach the Interface Card to the DB-25 Connector

Prepare Each End of the Ribbon Cable

§§§　Separate each end of the ribbon cable into five groups of wires by splitting the cable with the knife and pulling the groups of wires apart for about 50 millimeters.

✓　The red wire serves as a common ground.

✓　The single wire enables both latches' outputs.

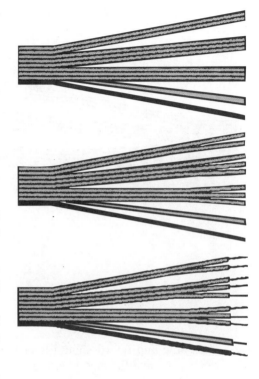

✓　The two-wire group enables either or both latches to capture data from the parallel port.

✓　The remaining two groups of three wires carry data (bits 0 through 5) from the parallel port.

§§§　Separate the wires in each group by splitting the cable with the knife and pulling the wires apart for about 20 millimeters.

§§§　Strip approximately 5 millimeters of insulation from the end of each wire at each end of the cable.

✓　Nick the plastic insulation with the knife, then pull off the insulation with your fingers.

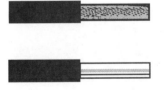

🖝　Heat each of the bare wires and "tin" it by saturating the wire's strands with solder.

🖝　Tin all wires at both ends of the ribbon cable.

Solder the Cable to the Interface Card

◉ Insert the wires into the component side of the board, matching numbered wires and holes.

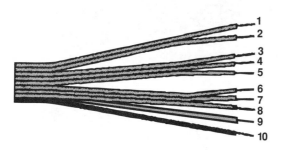

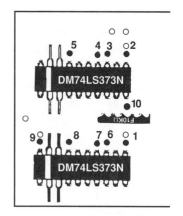

✓ The cable and board should look like this when you have finished inserting the wires into the interface card.

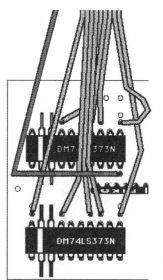

✓ Next, **carefully** turn the interface card over, keeping the wires in place on the interface card.

✓ The wiring side will be uppermost.

✎ Solder the cable wires to the pads on the wiring side of the interface card.

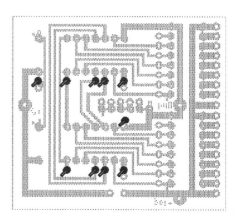

Solder the Cable to the DB-25 Connector

◉ Start by inserting one wire into the DB-25 connector, matching the numbered wire to the numbered solder socket as shown in the illustration.

✗ Do **not** match the cable wires to the tiny numbers on the DB-25 connector. Use the numbers in the illustration.

✍ Continue to insert and solder one wire at a time, in order from 1 to 10, until finished.

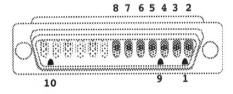

✓ The cable and DB-25 connector should look like this when you have finished inserting the wires into the interface card.

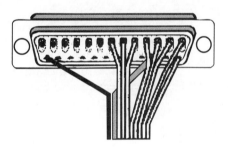

Solder Jumpers and Power Wires to the Board

$$$ Cut three jumper wires, each about 30 millimeters long.

$$$ Strip about 5 millimeters of insulation from each end of all the jumper wires.

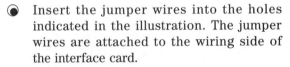

◉ Insert the jumper wires into the holes indicated in the illustration. The jumper wires are attached to the wiring side of the interface card.

✍ Solder the jumper wires to the pads on the interface card.

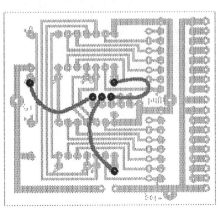

§§§   Cut two power leads, each about 70 millimeters long.

§§§   Strip about 5 millimeters of insulation from each end of each power lead.

◉   Insert the power leads into the holes indicated in the illustration. The power leads are attached to the wiring side of the interface card.

✍   Solder the power leads to the pads on the interface card.

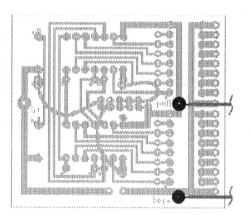

 You can use the same power supply to drive the nitinol and power the latches (approximately 4.5 volts and 1.5 amps) or you may use a separate power supply for the latches and the nitinol. **Choose now**, then follow the appropriate instructions to either **jumper the interface card or add a third power lead.** Of course, you can always change your mind—and the wiring—later!

To use the same power supply for the nitinol drivers and the latches:

§§§   Cut one jumper wire about 20 millimeters long.

§§§   Strip about 5 millimeters of insulation from each end of the jumper wire.

◉   Insert the jumper wire into the holes indicated in the illustration. The jumper wire is attached to the wiring side of the interface card.

✍   Solder the jumper wire to the pads on the interface card.

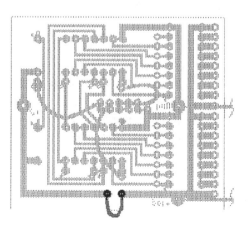

To use different power supplies for the nitinol drivers and the latches:

§§§   Cut one power lead about 70 millimeters long.

§§§   Strip about 5 millimeters of insulation from each end of the power lead.

◉   Insert the power lead into the hole indicated in the illustration. The power lead is attached to the wiring side of the interface card.

🖎   Solder the power lead to the pad on the interface card.

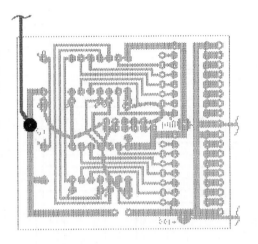

CHECK THE BOARD

✓   Test adjacent pads with a volt-ohmmeter to make sure they are not shorted.

✓   Check continuity between inputs and outputs with a volt-ohmmeter to be sure that solder joints are present and well-made.

✓   Visually inspect the board to make sure that the integrated circuits and the jumper wires have been installed correctly.

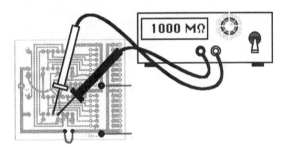

✓   Connect leg+ to a +5 volt power source. Connect ground on the interface card (gnd) to ground on the power supply. Turn the power supply on.

✗   The board should **not** emit smoke, flame, or sparks.

✓   Check that the DS2001 outputs are either at 4.5 volts or ground. If so, it's time to program the board.

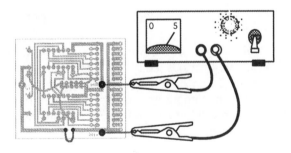

Interface Card Schematic

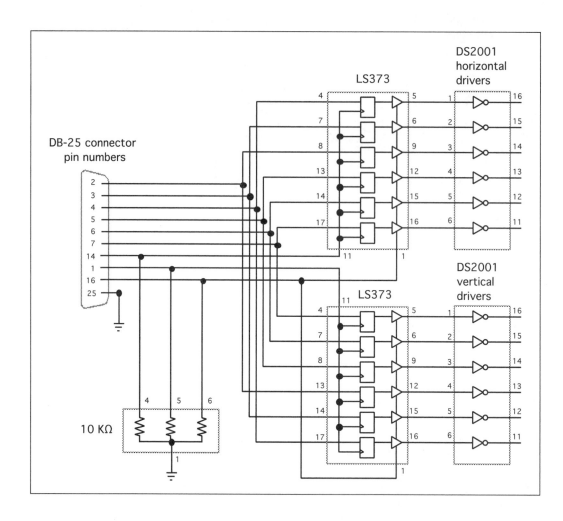

CONTROLLING STIQUITO II AND TENSIPEDE

In this section you will learn how to program Stiquito II and Tensipede, and make them walk using various gaits. The sample programs use the parallel interface as an open-loop system. In an open-loop system the robot has no sensors to provide feedback to the controller. Without any feedback the controller cannot vary its commands if an actuator fails or the robot gets stuck. For simple experiments, this is an adequate approach. More complex behaviors require sensors and feedback, and are outside the scope of this chapter.

Attach Stiquito II or Tensipede to the Interface

✓ The driver outputs to control your Stiquito II or Tensipede are located to the right of the DS2001 driver integrated circuits (as viewed with the board component side upward and pin 1 of the DS2001 driver oriented to the top left).

✓ Hook-up wires (magnet wire or wire-wrap wire) can be used to carry power and driver signals to Stiquito II or Tensipede.

✓ Six (6) vertical driver signals and six (6) horizontal driver signals are available.

✓ Two (2) spare driver signals are available, one on each DS2001 integrated circuit. A row of jumper points allows any of the six vertical or horizontal inputs to generate the spare driver signal.

✓ When you turn the interface card over to solder hook-up wires to the driver outputs, the signals are located as shown in the illustration.

✍ Solder hook-up wires directly to the board, or use a header strip (row of sockets) to connect the hook-up wires so that they may be easily inserted and removed.

✓ It is easiest to provide a power connection to each body segment of Stiquito II. There are four power pads on the interface card to provide enough connection points for power wires. All power is provided by a single supply.

✓ Note that the individual actuators are powered when a positive input to the DS2001 causes the actuator to be **connected to ground**. The power bus is **positive** and provides the current to heat the nitinol.

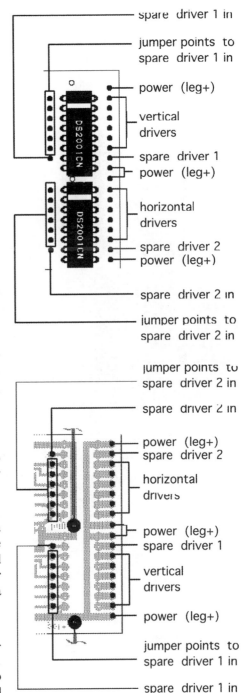

To use the programs in this chapter **without change** to control your Stiquito II, connect the wires between the interface card and Stiquito II **exactly** as shown in the following illustration. **Reversed text** (white on black background) indicates an **active low** signal. The pairs of numbers are the latch function and actuator identification for testing the robot.

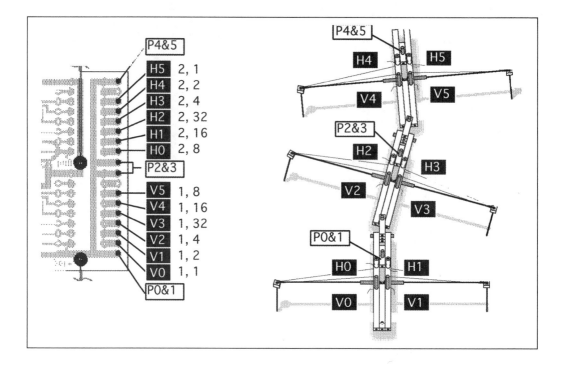

To use the programs in this chapter **without change** to control your Tensipede, connect the wires between the interface card and Tensipede **exactly** as shown in the following illustration. **Reversed text** indicates an **active low** signal. The pairs of numbers are the latch function and actuator identification for testing the robot.

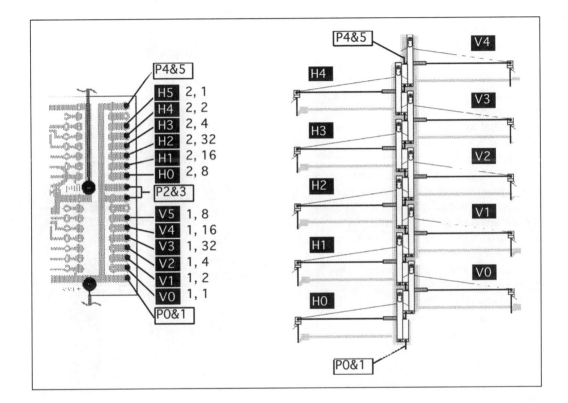

Programmer's Guide to the Interface

Overview—In IBM PC and compatible computers there is a parallel printer interface that has a **control port** and a **data port**. Data from the most recent write to the data port is held **inside the PC**, but to hold two bytes simultaneously, latches on the interface card must be used.

When the interface card is attached to the parallel printer port, the **control port** is used to choose whether the horizontal driver latch, the vertical driver latch, both latches, or neither latch will **track** or **capture and hold** data sent to the **data port**.

The DB-25 connector is used to connect the interface card to the parallel printer port. Data and control signals are sent from the program to the control and data ports, then via the DB-25 connector through the ribbon cable to the interface card. The data are latched on the interface card, then, if the latches' output is enabled, passed to the drivers, which activate the corresponding nitinol actuators.

Only three of the six bits available in the control port are used by the interface card.

Only six of the eight bits available in the data port are used.

The six data bits are **doubled to 12 bits** by being captured in separate latches. Because each latch can be controlled individually, the data each latch holds can differ. This allows 10 actuators on a Tensipede or 12 actuators on a Stiquito II to be controlled.

The Control Port—The three bits of the control port determine the operation of each latch on the interface card. Use the control port to choose whether you want to pass data to the DS2001 drivers (bit 2), and whether you want each latch to track the data you send in subsequent writes to the data port, or capture and hold the data you sent in the previous write to the data port (bits 1 and 0). Note that when a latch does not pass or track data, it has captured and is holding constant the data from the **last time it tracked a data port write**, not the data port write that follows the control port write, and definitely not from the control port write. **A value written to the control port is not seen at all by the data port.**

This illustration summarizes the operation of the control port bits.

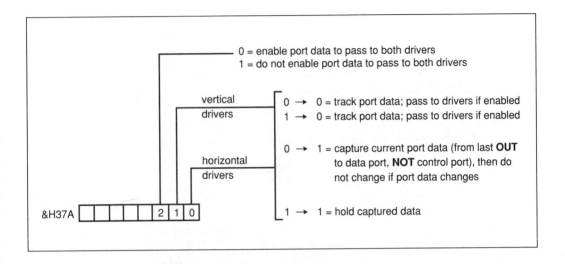

Typically the IBM PC and compatible computer control port is located at the **hexadecimal address 37A**, written in BASIC as **&H37A**. BASIC output statements for some useful functions of the interface card are listed below.

```
OUT  &H37A,1   'subsequent data port writes tracked & sent to vertical driver
OUT  &H37A,2   'subsequent data port writes tracked & sent to horizontal driver
               'the following commands do not pass data to the driver chips
OUT  &H37A,4   'subsequent data port writes tracked by both latches
OUT  &H37A,5   'previous data port write captured by horizontal latch
OUT  &H37A,6   'previous data port write captured by vertical latch
```

The relationship between the control port and the interface card is shown in the following figure. **Reversed text** indicates an **active low** signal.

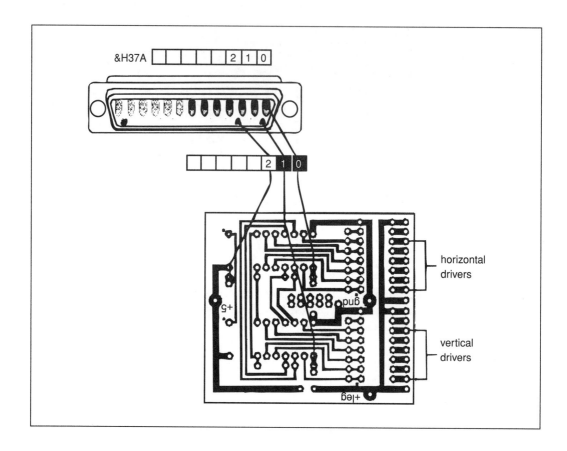

The Data Port—The data port for the IBM PC and compatible computer printer interface is located at the **hexadecimal address 378**, written in BASIC as **&H378**. A data port write is **usually** combined with a control port write to direct the data to the desired latch and its associated driver. Occasionally it is meaningful to write the same data to each latch, **usually** when both latches are initialized to zero. In this case only one control port write is necessary. The following BASIC program fragment illustrates the no-glitch approach to activating the horizontal and vertical drivers.

```
OUT &H378,V   'send vertical data to card
OUT &H37A,1   'turn vertical actuators on/off
OUT &H37A,3   'capture and hold vertical data
OUT &H378,H   'send horizontal data to card
OUT &H37A,2   'turn horizontal actuators on/off
OUT &H37A,3   'capture and hold horizontal data
```

The relationship between each bit in a byte sent to the data port, and the signals at the output of the drivers on the interface card (identified by the bit address), is summarized in the following figure. **Reversed text** indicates an **active low** signal.

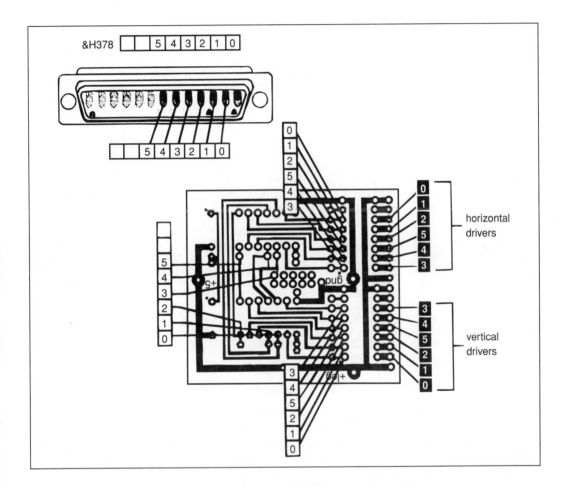

Driving Nitinol Actuators with a Constant Current—The program to drive the nitinol actuators shown in the previous section can be shortened because the nitinol actuators are not affected by very short changes in drive current. Here is the glitch-free driver program, and the timing diagram that corresponds to it. Notice that the actuators always have the desired data driving them, whether from the previous write to the horizontal or vertical drive latch or the current write. **Reversed text** indicates an **active low** signal.

```
OUT &H378,V        'send vertical data to card
OUT &H37A,1        'turn vertical actuators on/off
OUT &H37A,3        'capture and hold vertical data
OUT &H378,H        'send horizontal data to card
OUT &H37A,2        'turn horizontal actuators on/off
OUT &H37A,3        'capture and hold horizontal data
```

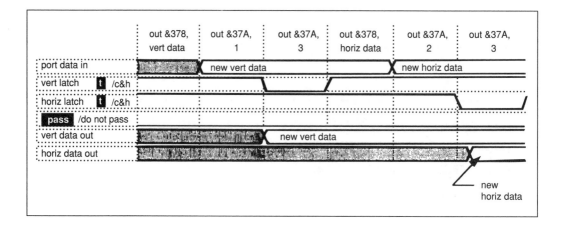

This program can be shortened if we are willing to accept a glitch, or unwanted data present for a short time, as the actuators are driven.

```
OUT &H378,V        'send vertical data to card
OUT &H37A,1        'turn vertical actuators on/off
OUT &H378,2        'capture and hold vertical data; turn horiz actuators
                   'on/off
OUT &H378,H        'send horizontal data to card
OUT &H37A,3        'capture and hold horizontal data
```

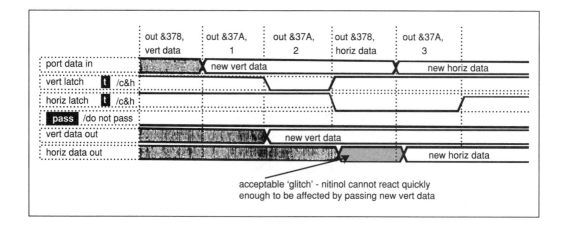

acceptable 'glitch' - nitinol cannot react quickly
enough to be affected by passing new vert data

The program can be shortened even further if we accept two glitches and can guarantee that the data written to the IBM PC parallel printer port will not change while the horizontal nitinol actuators are being driven. The second condition allows the printer port to drive the DS2001 chip directly: The 74LS373 latch passes the data until new data are written to the printer port, and never captures and holds data for the DS2001 horizontal driver chip.

```
OUT &H37A,1           'turn vertical actuators on/off
OUT &H378,V           'send vertical data to card
OUT &H378,2           'capture and hold vertical data; turn horiz actuators
                      'on/off
OUT &H378,H           'send horizontal data to card
```

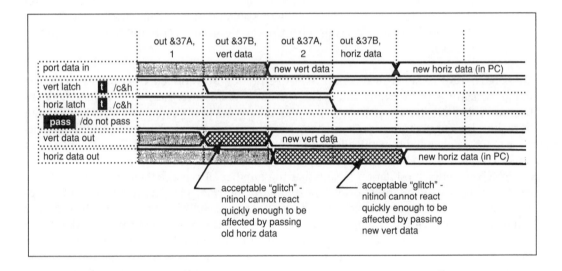

Because the nitinol reacts so slowly, the glitches are acceptable, as can be seen by this program fragment that runs on a 33MHz 486 laptop. The program must wait while a loop executes 9,000 times to ensure that the nitinol is fully contracted, and 3,000 times to allow the nitinol to expand.

```
OUT &H37A, 1          'contract vertical and horizontal actuators
OUT &H378, 49         'to move all legs for one step
OUT &H37A, 2
OUT &H378, 49
FOR x = 1 TO 9000
NEXT x

OUT &H37A, 0          'let the actuators expand for the next step
OUT &H378, 0
FOR x = 1 TO 3000
NEXT x
```

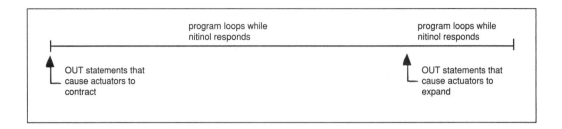

 The loop constant **must** be changed for different computers because they run BASIC at different speeds. Begin with a small loop constant and increase it until the nitinol moves. **Starting with the loop constant too high can damage the nitinol.**

Driving Nitinol Actuators with a Pulse-Frequency-Modulated Current—Driving the nitinol actuator with the same amount of current is unnecessary after the nitinol has contracted. Only enough current to keep the nitinol contracted is needed, which is just enough to replace the energy that escapes as heat. The current and the voltage supplied to the nitinol cannot be changed dynamically, but the **power** can be varied using a technique called **pulse frequency modulation (PFM)**.

Pulse frequency modulation means that the number of pulses (their frequency) is varied over time. The PC parallel printer port and the interface card can generate a PFM signal because the nitinol reacts slowly compared to the speed with which a BASIC program can turn the DS2001 driver chips on and off. By varying the length of time that the driver chips are left off, the frequency of the pulses can be increased or decreased. This allows the power used to drive the robot to be varied dynamically. The nitinol actuator behaves as a **leaky integrator** of the current pulses sent to it: The nitinol actuator responds to the heat generated by the current pulses and lost to convection from the wire. The following figures show how a PFM driver program works.

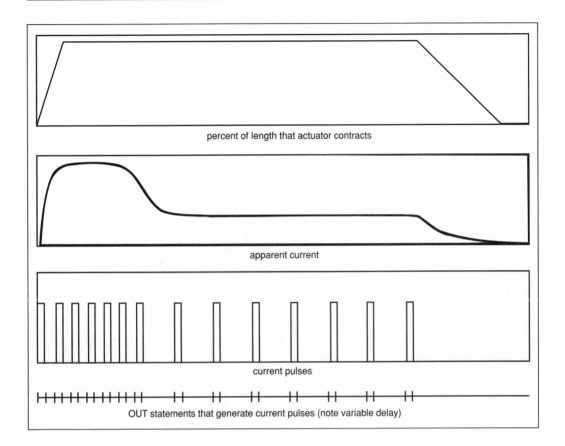

percent of length that actuator contracts

apparent current

current pulses

OUT statements that generate current pulses (note variable delay)

The following BASIC program subroutines illustrate how a PFM driver is constructed.

```
                    'contract vertical and horizontal actuators
                    'to move all legs for one step

OUT &H37A, 0        'pass all data port writes to both drivers (this
                    'works only if a tripod gait is being used)

FOR x = 1 to 20
GOSUB 100           'high frequency pulses initially contract actuators
NEXT x

FOR x = 1 to 40
GOSUB 200           'low frequency pulses maintain actuator contraction
NEXT X

OUT &H378, 0        'let the actuators expand for the next step
FOR x = 1 TO 2000
NEXT x
END

                    '-----------------
100                 'high frequency pulses
OUT &H378, 49       'drivers on
FOR x = 1 TO 50
NEXT x

OUT &H378, 0        'drivers off
FOR x = 1 TO 50
NEXT x

RETURN

                    '-----------------
200                 'low frequency pulses
OUT &H378, 49       'drivers on
FOR x = 1 TO 50
NEXT x

OUT &H378, 0        'drivers off
FOR x = 1 TO 100
NEXT x

RETURN
```

Introduction to Gaits with Sample Programs

Gaits—The mechanisms of arthropod loco-motion are complex and have been studied extensively. The structure of an insect leg is also quite complicated. But, even though Stiquito II and Tensipede are simple, the fundamental features of arthropod locomotion can be demonstrated by small programs. Later, if you choose, you can develop more realistic models of gait controllers based on neural networks or central pattern generators, and feedback from strain gauges or other sensors that mimic the sensorimotor loop in a real insect.

The gaits of insects are believed to be due to central pattern generators that vary the animal's gait from a **metachronal wave** to a **tripod gait**, and all the variations in between. Each gait conserves energy as it preserves the balance of the insect. As the animation sequences indicate, the insect is always in a stable position with at least three legs on the ground at all times.

The metachronal wave is the slowest and most stable gait. It is seen when a "wave" of leg movement ripples down each side of the insect or arthropod. The animation sequence shows two "waves" flowing down each side of the Tensipede. The tripod gait is the fastest stable gait, with two legs on one side of the insect and one on the other side alternately on the ground or in the air, as shown in an animation of Stiquito II.

These two extremes in locomotion can be programmed into Stiquito II and Tensipede. Central pattern generator controllers are left as an exercise for the reader; they are covered in detail in books by Donner and Beer.

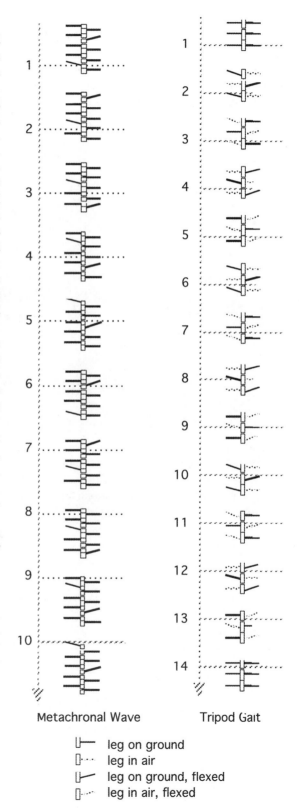

Metachronal Wave   Tripod Gait

⊢— leg on ground
◻··· leg in air
⊢╱ leg on ground, flexed
◻··· leg in air, flexed

Programming the Metachronal Wave for Tensipede—Before you enter and run the Tensipede metachronal wave program, use the following program to verify that the actuator wires are correctly hooked up. The illustration shows a pair of numbers next to each actuator. These numbers are read as **latch function, actuator control number** for each leg, and repeatedly entered into the BASIC program when requested until you are satisfied that all actuators work correctly. You can activate multiple actuators simultaneously by entering the sum of their actuator control numbers. To make the Tensipede legs designated by H4 and H1 move together, enter 2 + 16, or 18, as the actuator control number.

```
10  OUT &H378, 0 'init both latches
    OUT &H37A, 4

    INPUT "latch function"; L
    INPUT "actuator control number"; A

    OUT &H37A, L
    OUT &H378, A

    FOR x = 1 TO 9000: NEXT x
    GOTO 10
```

The following program causes Tensipede to walk forward. Because the robot is not articulated, it does not turn or negotiate obstacles well. If the robot just twitches and does not move forward, bend the tip of each leg backward to make a **ratchet foot**. This will allow the driven leg to catch the surface on which it is standing, pushing the rest of the legs forward along the backward-slanted foot.

```
    OUT &H37A, 0     'initialize both latches and all legs
    OUT &H378, 0

3   PRINT : PRINT : PRINT "turn on power supply now, then..."
    PRINT "press any key to start tensipede walking"

5   start$ = INKEY$
    IF start$<>"" THEN GOTO 7 ELSE GOTO 5

7   PRINT "press any key to stop tensipede"
    PRINT "(you may have to wait a second or two)"

10  OUT &H37A, 1
    OUT &H378, 32
    OUT &H37A, 2
    OUT &H378, 8
    FOR x = 1 TO 9000: NEXT x
```

```
OUT &H37A, 0
OUT &H378, 0
FOR x = 1 TO 3000: NEXT x

OUT &H37A, 1
OUT &H378, 16
OUT &H37A, 2
OUT &H378, 16
FOR x = 1 TO 9000: NEXT x

OUT &H37A, 0
OUT &H378, 0
FOR x = 1 TO 3000: NEXT x

OUT &H37A, 1
OUT &H378, 1
OUT &H37A, 2
OUT &H378, 32
FOR x = 1 TO 9000: NEXT x

OUT &H37A, 0
OUT &H378, 0
FOR x = 1 TO 3000: NEXT x

OUT &H37A, 1
OUT &H378, 2
OUT &H37A, 2
OUT &H378, 4
FOR x = 1 TO 9000: NEXT x

OUT &H37A, 0
OUT &H378, 0
FOR x = 1 TO 3000: NEXT x

OUT &H37A, 1
OUT &H378, 4
OUT &H37A, 2
OUT &H378, 2
FOR x = 1 TO 9000: NEXT x

OUT &H37A, 0
OUT &H378, 0
FOR x = 1 TO 3000: NEXT x

a$ = INKEY$

IF a$<>"" THEN OUT &H37A, 5: OUT &H378, 0: GOTO 3

GOTO 10
```

Programming the Tripod Gait for Stiquito II—Before you enter and run the Stiquito II tripod gait program, use the following program to verify that the actuator wires are correctly hooked up. The illustration shows a pair of numbers next to each actuator. These numbers are read as **latch function, actuator control number** for each leg, and repeatedly entered into the BASIC program when requested, until you are satisfied that all actuators work correctly. You can activate multiple actuators simultaneously by entering the sum of their actuator control numbers. To make the Stiquito legs designated by V4 and V1 lift at the same time, enter $2 + 16$, or 18, as the actuator control number.

```
10  OUT &H378, 0 'init both latches
    OUT &H37A, 4

    INPUT "latch function"; L
    INPUT "actuator control number"; A

    OUT &H37A, L
    OUT &H378, A

    FOR x = 1 TO 9000: NEXT x
    GOTO 10
```

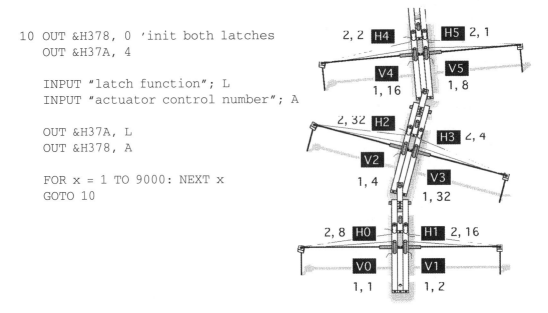

The following is a program to make Stiquito II walk forward using a tripod gait. Because the robot is articulated, you can place small obstacles in its path, and it will negotiate its way around or over them. Stiquito II will climb up and down a staircase with three or four broad shallow steps (about 3 inches deep by 8 inches wide by ⅛ inch high).

```
    OUT &H37A, 0     'initialize both latches and all legs
    OUT &H378, 0

3   PRINT : PRINT : PRINT "turn on power supply now, then..."
    PRINT "press any key to start stiquito walking"
5   start$ = INKEY$
    IF start$<>"" THEN GOTO 7 ELSE GOTO 5

7   PRINT "press any key to stop stiquito"
    PRINT "(you may have to wait a second or two)"

10  OUT &H37A, 1
    OUT &H378, 49
    OUT &H37A, 2
    OUT &H378, 49
    FOR x = 1 TO 9000
    NEXT x
```

```
OUT &H37A, 1
OUT &H378, 49
OUT &H37A, 2
OUT &H378, 0
FOR x = 1 TO 9000
NEXT x

OUT &H37A, 0
OUT &H378, 0
FOR x = 1 TO 3000
NEXT x

OUT &H37A, 1
OUT &H378, 14
OUT &H37A, 2
OUT &H378, 14
FOR x = 1 TO 9000
NEXT x

OUT &H37A, 1
OUT &H378, 14
OUT &H37A, 2
OUT &H378, 0
FOR x = 1 TO 9000
NEXT x

OUT &H37A, 0
OUT &H378, 0
FOR x = 1 TO 3000
NEXT x

a$ = INKEY$

IF a$<>"" THEN OUT &H37A, 5: OUT &H378, 0: GOTO 3

GOTO 10
```

Conclusion: Some Projects and Experiments

Here are some experiments and projects to try. Some are very difficult, others are very easy.

- Attach different feet to walk on various surfaces; use the results to design a dynamically reconfigurable foot actuated by nitinol.
- Use micro strain gauges to provide feedback to the controller about leg position.
- Compute the Froude number of Stiquito II, Tensipede, and any insects you can obtain; compare the numbers to determine how insectlike nitinol-propelled robots are.

- Investigate different gaits and different ways to implement the tripod gait with Stiquito II; try to discover the optimal gait for Stiquito II and prove that it is or is not the tripod gait, and that it is or is not the same as an ant's actual gait (you may want to videotape both an ant and Stiquito II as they walk for later analysis).
- Design and build a self-contained Stiquito II or Tensipede that uses sensors, a microprocessor controller, and batteries to operate autonomously.
- Design and build a Stiquito II colony to investigate if and how cooperative and competitive behavior can emerge from primitive goals programmed into each robot.
- Design a genetic algorithm to evolve the tripod gait in Stiquito II.
- Build a solar-powered Stiquito II or Tensipede.
- Design a "space Stiquito" that uses shadow panels to alternately place nitinol actuators in light and shade; this is a mechanical oscillator with the side-effect of causing the robot to walk.
- Develop a gait controller that can be varied continuously from a metachronal wave gait to a tripod gait by changing only one variable.
- Use analog VLSI retina chips to control locomotion.
- Design and build new effectors using nitinol.
- Design and build sensors for heat, light, sound, motion, toxic chemicals, or radioactivity.
- Build a snake using the basic leg as a rib. Program the snake for serpentine locomotion.
- Build a biped using the basic leg as a segment in a larger biped leg.
- Store nitinol's energy to build a robot flea. Find an analog to the substance *resilin* used by fleas to incrementally accumulate energy for their leaps.

Stiquito II and Tensipede are platforms for research and education. There are many more things to do with them than I have listed here. Build one of these robots yourself, and I hope you will get an idea for a Stiquito II or Tensipede project that interests you. If you do, then you will understand why I have taken the time, and tried my best, to teach you how to build these robots.

ACKNOWLEDGMENTS

I want to thank all the persons who have written to suggest improvements and tell me of their work with Stiquito. There are hundreds of you who have found different and interesting ways to use this tiny robot. If you had not showed such interest in Stiquito, I would not have written this book, or at least not invested as much effort in it.

Jon Blow, a graduate student in programming language design at the Experimental Computing Facility at the University of California, Berkeley, established and maintains the Stiquito mailing list that provides a forum for Stiquito users. Send requests to be added to or removed from the list to

    stiquito-request@xcf.berkeley.edu

Special thanks to the Stiquito II beta testers.

    John Bay, Virginia Polytechnical Institute (Blacksburg)
    Jim Conrad, BPM Technology
    John Estell, Bluffton College
    Russ Fish, University of Utah
    Deryl Shields, Boeing Aerospace

Rod Douglas, Christof Koch, and Terry Sejnowski, organizers of the NSF Neuromorphic Engineering Workshop, let me try Stiquito II out on a group of VLSI designers. I observed many mistakes that occur while building Stiquito II. A number of participants built working robots . . . well, almost-working robots.

All mistakes and errors in this chapter are mine. If, however, you do not read this chapter before you try to build the robot, the mistakes and errors you make will belong to you alone.

REFERENCES

References are legion because robotics spans so many disciplines. These references have been selected because they provide some jumping-off points for research using Stiquito II, not a survey of all relevant books and articles.

Alexander, R.M. 1992. *Exploring biomechanics.* New York: Scientific American Library. W. H. Freeman and Company.

Beer, R. 1990. *Intelligence as adaptive behavior.* Boston: Academic Press.

Brooks, R. 1990. A robot that walks: Emergent behaviors from a carefully evolved network. *Neural Computation* 1(2): 253–262.

Donner, M. 1987. *Real-time control of walking.* Cambridge, Mass.: Birkhauser Boston.

Gilbertson, R.G. 1992. *Working with shape memory wires.* San Leandro, Calif.: Mondo-Tronics, Inc.

Mead, C. 1989. *Analog VLSI and neural systems.* VLSI System Series/Computation and Neural System Series, L. Conway, C. Seitz, and C. Koch, eds. Reading, Mass.: Addison-Wesley.

Mills, J. 1992. Area-efficient implication circuits for very dense Łukasiewicz logic arrays. 22nd Int'l Symp. Multiple-Valued Logic. Sendai, Japan. pp. 291–299.

Chapter 7

An M68HC11 Microcontroller-Based Stiquito Controller

James M. Conrad and Mohan Nanjundan

An autonomous Motorola M68HC11-based controller was designed for the low-cost, legged robot Stiquito to control its walking motion. The main goals were to enhance the functionality of the robot and to give it autonomy. Since Stiquito is not capable of carrying much weight, simple infrared sensor circuitry was interfaced with the microcontroller. A handheld infrared remote transmitter was also designed. Stiquito is able to execute commands sent through the transmitter and respond appropriately. This basic design can be easily modified in the future for more sophisticated control of Stiquito-like robots.

INTRODUCTION

Legged locomotion is becoming a popular area of research. Since the advent of powerful, single-chip microcomputers in the late 1980s, the advances made in legged robotics have been noteworthy. The hardware used to control the walking motion of earlier robots was complex;[1] but with the advent of microcomputers on chip, the electronics required to perform many difficult tasks have been reduced considerably.

An autonomous controller, based on the Motorola M68HC11 microcontroller, was designed to control the walking motion of Stiquito. The M68HC11-based autonomous motion controller performs two functions.

1. Receiving and decoding commands from an infrared transmitter

2. Taking appropriate control action for the hexapod robot to move

An infrared transmitter sends specific codes. The infrared receiver translates the code into logic levels, which are decoded by the microcontroller. The microcontroller activates the appropriate legs of the robot corresponding to the code sent through the driver circuit. The basic block diagram of the autonomous motion controller is shown in Figure 7.1.

152

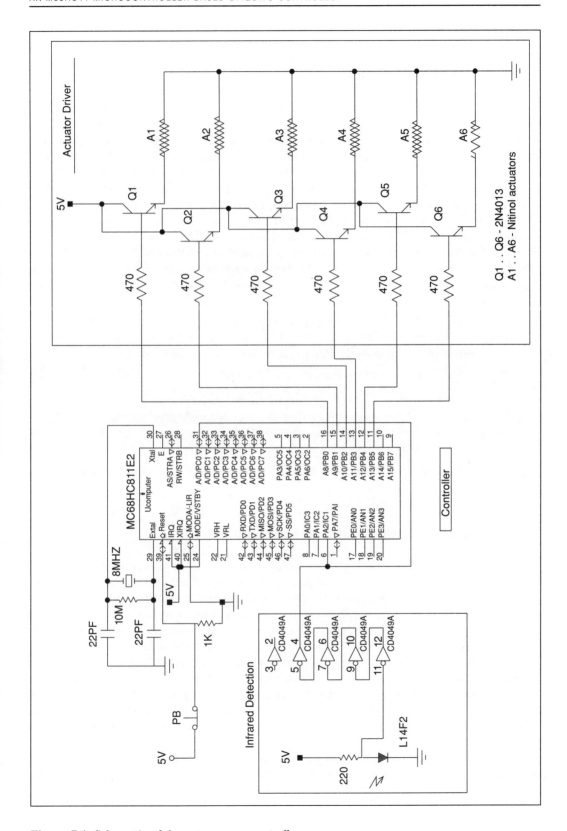

Figure 7.1. Schematic of the autonomous controller.

CONTROLLER CIRCUITRY

The M68HC11 Microcontroller

The Motorola MC68HC811E2 microcontroller is a versatile single-chip microcomputer with many peripherals built on a single chip[2]. The on-chip peripherals include an eight-channel analog-to-digital converter, an asynchronous serial communications interface, and a separate synchronous serial peripheral interface. In addition, the main 16-bit free-running timer has three input-capture lines, five output-compare lines, and a real-time interrupt function. The input-capture and output-compare lines produce software interrupts. An 8-bit pulse accumulator subsystem can count external events or measure external periods. The microcontroller also has on-board EEPROM (electrically erasable programmable read-only memory) of 2 kilobytes for program code. On-board random-access memory (RAM) of 256 bytes is provided for temporary storage of program variables. The microcontroller has an elaborate input-output (I/O) system, where various pins are configured as either inputs or outputs under software control. There are five ports associated with the microcontroller.[3]

- Port A. All eight pins can be configured for general-purpose I/O or for timer or pulse accumulator functions. Pins 0, 1, 2 and 3, 4, 5 can be used as input-capture and output-compare pins, respectively. Pin 7 can be used as a pulse accumulator.

- Port B. All eight pins of this port are for output only.

- Port C. All eight pins of this port can be configured for bidirectional I/O.

- Port D. There are five pins in this port, which can be configured for bidirectional I/O. Port pins 0 and 1 could be used as receive (RxD) and transmit (TxD) pins if serial transmission of data is involved.

- Port E. All eight pins of this port are used as digital inputs or inputs to the on-chip analog-to-digital converter.

M68HC11 Microcontroller-Based Motion Controller

The circuit schematic of the M68HC11-based motion controller is shown in Figure 7.2. The microcontroller is operated in the single-chip mode. The MC68HC811E2 microcontroller has sufficient (2 kilobytes) EEPROM for the Stiquito motion controlling code. The interrupt pins (XIRQ and IRQ) have been disabled by connecting them to the source voltage. The RESET pin is operated manually; that is, after power-on, the reset is made low momentarily and then made high. The crystal (8 megahertz), capacitors, and resistor are connected between the crystal oscillator and clock pins (XTAL and EXTAL) as shown. The output from the infrared detector is connected to the input-capture 1 (IC1) pin of the microcontroller. This is also connected to bit 0 of port C. Port C bits 0 through 5 are connected to the base of NPN transistors. These port lines drive the base of the transistors. When the transistors turn on they connect the supply voltage (5 volts) to the actuator wires of the hexapod robot (Stiquito). This causes current to flow through these wires and heat them. When these wires heat they contract, pulling the legs to which they are connected, thereby causing them to move. By alternately turning the transistors on and off the motion of the robot is controlled through software.

A printed circuit board was specifically designed for this circuit, which incorporates the M68HC11 microcontroller, driver circuitry for the actuators, and a simple infrared detector circuit. The whole setup was mounted on the back of the Stiquito, along with a 6-volt battery pack to supply power to the controller and the actuator driver circuitry.

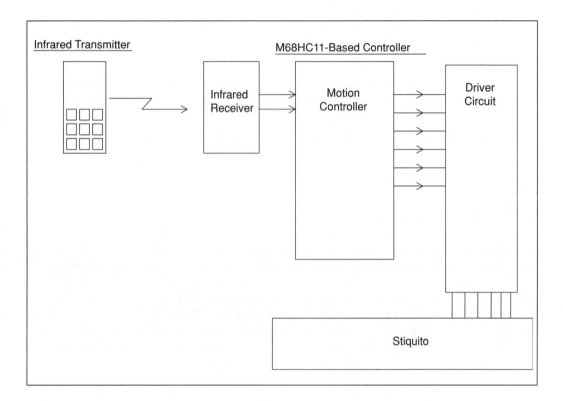

Figure 7.2. Block diagram of the M68HC11-based Stiquito autonomous controller.

Programming the MC68HC811E2

The M68HC11 Evaluation Module (EVM) was used to program the MC68HC811E2. The EVM programs a dual in-line pin (DIP) or a plastic-leaded chip carrier (PLCC) microcontroller. For the Stiquito a DIP MC68HC811E2 was used. The EVM is connected to an IBM PC through the serial port. A driver program EVM11 establishes communication between the IBM PC and the EVM. On compiling the source code, the assembler produces an .S19 file, which is downloaded to the microcontroller. The LOAD command of the driver program downloads the .S19 file to the EVM memory. The microcontroller is then programmed using the PROG command.

After the MC68HC811E2 was programmed, it was inserted into the target board and powered up. The microcontroller started operating as soon as it was reset. Since the program code was not complex, light-emitting diodes (LEDs) were switched on at certain parts of the code, providing visual information regarding the status of the microcontroller.

Advantages of the Microcontroller-Based Design

There are many advantages of the microcontroller-based design, including the following:

- Flexible control of the robot's motion can be obtained through software changes or by having a programming keyboard at the transmitter end. The present design allows selection of five preset walking modes.

- Interrobotic communications can easily be designed by adding transmitter circuitry to the controller board. This might help in studying emergent cooperative behavior in a colony of Stiquitos.

- Various sensors can be mounted (for example, ultrasonic, proximity, and so on) and interfaced to the microcontroller. Software code incorporated into the microcontroller would interface with the sensors.

- Complex gaits for the locomotion of Stiquito can easily be programmed.

INFRARED REMOTE CONTROLLER

Infrared Remote Control Unit

The infrared remote control unit is composed of the infrared transmitter and the infrared detector (receiver). The handheld transmitter (Figure 7.3 shows a schematic) is wired on a general-purpose printed circuit board and operates with a 6-volt battery pack. A Motorola MC14497 remote-control transmitter chip is hardwired for use with up to 32 keys, amplitude modulation (AM), and a logic 1 start bit[4]. The MC14497 is a complementary metal-oxide semiconductor (CMOS) biphase pulse-code modulated (PCM) remote-control transmitter chip in an 18-pin package. In standby mode it draws only 10 microamperes and operates within a range of 4 to 10 volts. Transmission and internal timing are controlled with a 500-kilohertz ceramic resonator.

The basic configuration of the MC14497 will support 32 keys. With two additional diodes and switches between pins 3-6 and 3-5, the capacity can be increased to a maximum of 62 keys[5]. The infrared LED is driven by an NPN transistor as shown in Figure 7.2. Pin 8 (output) of the MC14497 drives the base of the transistor[6].

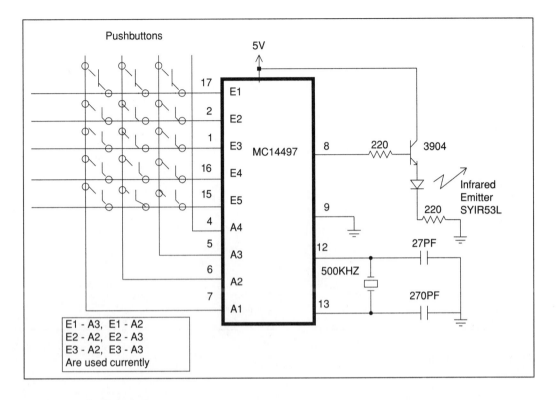

Figure 7.3. Schematic of the infrared transmitter.

Pulse Code Modulation

Biphase PCM signals are relatively easy to read if interpreted correctly. The most important part in reading PCM is keeping track of the bit times and noticing at what point the 0-to-1 or 1-to-0 logic transition occurs. If the pulse burst is sent or received during the first half of the bit time, the bit is a logic 1. Conversely, if the pulse occurs during the second half of the bit time, the bit is a logic 0.

When a key is pressed, the transmitter sends an automatic gain control (AGC) burst lasting a half-bit time, a start bit (logic 1), and a 6-bit PCM data word. The purpose of the AGC pulse preceding the PCM data is to set up the AGC loop in the receiver in time for the start bit. The 6 data bits are designated A (least significant bit) through F (most significant bit) and are shifted in, least significant bit first. While each bit is represented as a logic 0 or logic 1 level pulse having a duration of 0.5 or 1 millisecond, the actual output of the transmitter is a 40-kilohertz pulse burst for the duration of any logic 1 level. Only after the data are conditioned by the receiver will they appear as discrete logic levels. The transmitter repeatedly transmits the same code as long as the key is pressed. When the key is released, a channel 62 end-of-transmission code is automatically sent[7]. PCM coding is shown in Figure 7.4.

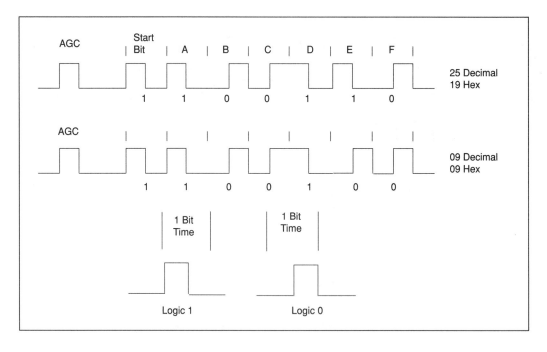

Figure 7.4. Pulse code modulation.

Infrared Detector

The Motorola MC3373 wideband amplifier-detector chip is designed for use with infrared pulse-burst transmissions. The entire receiver circuit shown is a two-chip envelope detector and transistor-transistor logic (TTL) level shifter. A photodiode receives the infrared pulses from the transmitter and amplifies them. When an infrared signal of approximately 40 kilohertz is perceived, the output goes low. Q1 inverts this signal and applies it to a series of 4049 CMOS inverters that are capable of driving the low-power Schottky TTL input load of the computer and lighting an additional LED so one can see that data are being received[8].

The infrared detector circuit should be used for long-range operation, but it is not necessary for short-range operation. For Stiquito, the remote-control unit was connected for short-range operation. Therefore, the receiver circuit was modified as shown in Figure 7.1. The infrared detector diode did the job of rectifying the 40-kilohertz carrier signal from the transmitter. This worked reliably for short ranges up to 2 feet. The main reason for this arrangement is to reduce as much as possible the number of components on the hexapod robot.

PCM Detection in Software

Since detection of the PCM data involves keeping track of the bit timing, translation can easily be accomplished using the M68HC11 microcontroller. The steps involved in the detection of the PCM data are as follows:

1. Set up the input-capture interrupt to detect falling edges. The output of the receiver is connected to the IC1 pin of the M68HC11 controller to trigger the input-capture interrupt, and is connected to bit 0 of port C to read the incoming PCM data.

2. When infrared data are received, the input-capture interrupt is invoked at the falling edge of the AGC pulse.

3. In the input-capture interrupt routine, the processor waits for the start bit, which occurs exactly 2 milliseconds after the AGC pulse.

4. After the start bit is received, the microcontroller checks for the logic level of bit 0 of port C every millisecond, since the data bits coming in after the start bit are 1 millisecond apart. If the logic level is 0, then the value of the data bit is 0; if the logic level is 1, the value of the data bit is 1.

5. As the data are being detected, the bit level detected at bit 0 of port C is rotated right into a memory location. Steps 4 and 5 are repeated until all the data bits are received.

6. After all six data bits have been received, the memory location where the data are stored is rotated twice right to pad the most significant bits with 0's.

SOFTWARE

Software Description

The software was written in M68HC11 assembly language. It consists of three parts.

1. The main routine, where the hexapod robot is made to walk in a particular mode (walk, pace, or caterpillar), depending on the code received from the remote-control unit.

2. The input-capture interrupt routine, which is invoked at the occurrence of the AGC pulse. The input-capture routine then decodes the infrared data occurring at bit 0 of port C.

3. The timer-interrupt routine, which keeps track of the timing for switching on each leg in the case of walk mode, and switching groups of legs in the case of the other modes of operation (pace, tripod, and so on).

The Main Routine

In the main routine, the initialization of the various interrupts is performed first. The program then checks the current mode of operation and jumps to the appropriate mode subroutine. In these subroutines the actuators are switched sequentially in the case of walk mode, or in groups in the case of the other modes. The number of modes that the Stiquito is made to walk is limited only by the number of different codes that can be sent by the transmitter: 32 in normal mode; 62 in enhanced mode. Currently, Stiquito is programmed to operate in five different modes.

Mode 1, walk. Each leg is activated individually.

Mode 2, pace. Groups of legs (three at a time) are moved alternately.

Mode 3, turn right. One group of legs is moved to make the robot turn right.

Mode 4, turn left. One group of legs is moved to make the robot turn left.

Mode 5, fast pace. Same as mode 2, but at a faster rate.

Additional modes can easily be incorporated in the future by programming new gaits corresponding to new transmitted codes.

The Input-Capture Routine

The IC1 interrupt routine is invoked at the falling edge of the AGC pulse. Once the program goes into the input-capture routine it disables all interrupts and keeps checking bit 0 of port C for the start bit. Once the start bit is sensed, a delay of about 0.5 milliseconds is introduced to position a software pointer at the middle of the start pulse. The start bit is discarded and a delay of 1 millisecond is introduced to position the pointer at the middle of the first data bit. This bit is then rotated right into a register (memory location). This process is repeated six times to acquire the six data bits. The register is then rotated twice to pad the most significant bits of the mode register with 0's. This mode byte is then checked with an end-of-transmission (EOT) code. This code is sent by the transmitter every time the pressed key is released. If the code sent is EOT, it is discarded. Thereafter, it returns from the interrupt routine. The main program takes care of the appropriate mode.

The Timer-Interrupt Routine

The timer-interrupt routine is invoked every 0.5 seconds. This is accomplished by prescaling the timer by 16 during initialization. A software counter is decremented in the timer-interrupt routine. This software counter value can be modified to obtain different timing for actuating the legs. The counter is currently loaded with 3, which gives 1.5-second timing. When this timer times out in the interrupt routine, it updates a leg counter and a toggle register. The leg counter gives the number of the leg to be activated in the case of walk mode. The other register, which is toggled every 1.5 seconds, gives information regarding which group of legs is to be activated in the case of the other modes, where groups of legs are activated according to the different gaits of operation.

CONCLUSIONS

The main objective of this experiment was to enhance the functionality of the motion controller for the Stiquito hexapod robot. Keeping in mind the size and the weight-carrying capability of the original Stiquito design, the Motorola MC68HC811E2 microcontroller was selected for this application. Even though the MC68HC811E2 microcontroller's built-in features were not fully exploited, the need will arise in future designs for a more sophisticated controller to be designed.

The existing system consists of a handheld remote transmitter and an MC68HC811E2-based intelligent Stiquito motion controller mounted on the back of the robot. A printed circuit board was specifically designed to house the microcontroller, driver, and the infrared detector circuitry. A 6-volt battery pack powered both the microcontroller and the driver circuitry. The hexapod robot successfully responded to commands sent from the infrared remote transmitter.

The hardware and software described forms the base for future microcontroller projects. The MC68HC811E2 is capable of controlling more legs, so it can be used to control a Stiquito II of the original Stiquito with two degrees of freedom. An RS-232C remote serial communication link between a PC and the controller will allow the movement of a colony of Stiquitos.

ACKNOWLEDGMENT

The authors would like to thank Motorola University Support for its donation of controller chips and manuals.

REFERENCES

1. Todd, D.J. 1985. *Walking machines: An introduction to legged robots.* London, U.K.: Kogan Page Ltd., 9–28.

2. Motorola, Inc. 1991. *M68HC11 reference manual.*

3. Ibid.

4. Ciarcia, S. 1987. Build an infrared remote controller. *Byte,* Feb.: 101–110.

5. Ibid.

6. Ibid.

7. Ibid.

8. Ibid.

Chapter 8

An M68HC11-Based Stiquito Colony Communication System

James M. Conrad, Gregory Lee Evans, and Joyce Ann Binam

An autonomous M68HC11-based controller was designed with the ability to send and receive infrared signals. This design was an improvement of an existing model; the previous controller received an infrared signal that was used to control the movement of an attached Stiquito robot. The infrared signal was generated with a small infrared remote transmitter.

The new design incorporates the idea of a colony containing a queen and several colonists. The queen generates movement commands by using photocells to detect directional light levels, generating and transmitting infrared movement signals based on which photocells have detected light. The infrared signal is received by the colonists, the appropriate movement is executed, and the signal is retransmitted to other colonists. Leg movements are displayed using six light-emitting diodes (LEDs) on the controller. This design can be enhanced in the future to incorporate new areas of interest.

INTRODUCTION

Researchers are continually making advances in the robotics field. An advancement that can further expand the capability of robots is to design robots that work together in colonies. The limiting factor in this process is communication.

Even without communications, robots can work together in highly predictable environments. Research has shown, however, that robots work better and can be used in a wider variety of situations when they are able to communicate with each other[1]. Therefore, the aim of this project was to further the current research in the area of interrobotic communications.

For this project, a communications system was designed that could be applied to a colony of autonomous robots. The system was specifically designed for use with microcontrolled Stiquito robots[2]. Using the original Stiquito design, an M68HC11 microcontroller-based Stiquito[3] was constructed. The microcontroller-based design receives infrared signals and, based on those signals, controls the leg movements of the attached Stiquito robot. A hand-held infrared remote is used to generate the leg movement signals. Five Stiquito walking modes were incorporated into that design: walk, left turn, right turn, pace, and fast pace.

This M68HC11-based Stiquito controller was used as a starting point for the communication system project. A colony of controllers, consisting of one queen controller and an unlimited number of colonist controllers, was generated from the original stand-alone microcontroller design. The colony was designed to need little outside interaction. The queen controller generates movement commands in response to lighting conditions. Light can be interpreted as being received from the left, the right, or the front. If the queen controller receives adequate light from any of the three directions, she will detect it and generate an appropriate movement command to be passed to the colonists. The colonists receive the movement commands, act on them accordingly, and pass them on to other colonists.

PROJECT BACKGROUND

The colony communication system project represents third-generation research. The communication system was designed as an advancement for an existing microcontrolled Stiquito robot, which was built based on the original Stiquito robot design.

The microcontroller-based Stiquito was developed at the University of Arkansas by Mohan Nanjundan and James Conrad. They designed an M68HC11-based controller that attaches to the top of a Stiquito robot and manipulates its leg movements. An MC68HC811E2 microcontroller chip is used in the design. The mounted controller can receive an infrared signal, decode it, and move the legs of the Stiquito accordingly. Five walking modes were incorporated into the microcontroller design: walk, left, right, pace, and fast pace. A handheld remote control based on a design by Ciarcia[4] was built to generate and transmit the signals for the five walking modes. Figure 8.1 shows a schematic of the controller that was designed.

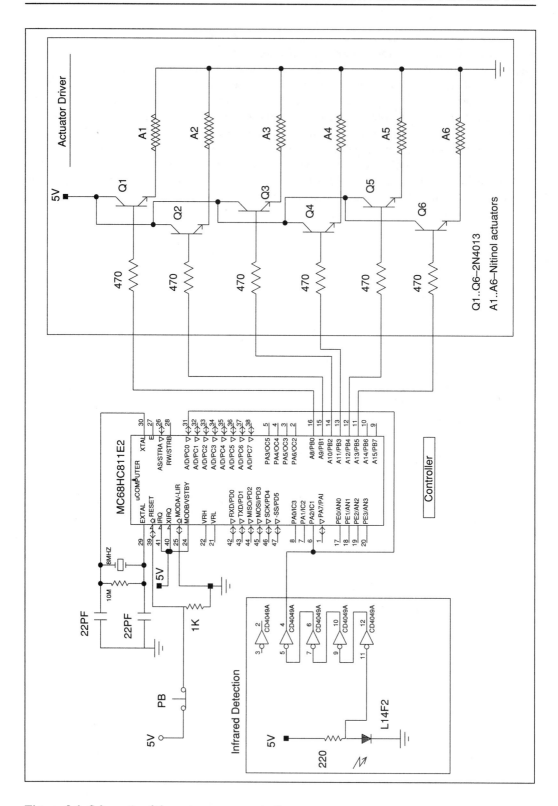

Figure 8.1. Schematic of the autonomous controller.

PROJECT DESCRIPTION AND PROCEDURES

Project Evolution

This project was completed by two computer systems engineering students for a senior design class at the University of Arkansas. The students' primary goal in choosing a project was to gain hardware experience. With this in mind, they began researching hardware-related projects. At the same time a professor in the computer systems engineering department, James Conrad, was conducting research in the area of autonomous robots. The microcontroller-based Stiquito was part of his research. He proposed that the students in some way enhance the existing microcontroller-based Stiquito design. The Stiquito project had a lot of potential for three main reasons. First, any research conducted in this area would likely help further research that was already being done at the university. Second, there were many spin-off areas and enhancement possibilities associated with the micro-controlled Stiquito. Third, working on an existing project provided an optimal way for the two students to gain hardware experience. They could gain basic experience by replicating the existing design, and they could advance by designing and implementing their own hardware.

Several project areas could be pursued; the students became interested in implementing a communication system for the Stiquito robot. A communication system would greatly enhance the current microcontrolled robot. It would allow multiple Stiquitos to work together, whereas the previous project had allowed for only one stand-alone robot. (More enhancements are discussed in the section Key Features and Functionality Improvements.)

Project Description

The design project consists of a colony of controllers. The colony is made up of two types of controllers: one queen controller and an unlimited number of colonist controllers. The queen is responsible for initiating all of the movement commands that are passed on to the colonists. She generates movement commands by interpreting light levels. She detects whether light is coming from the front, the left, or the right, and, based on the direction of the light, generates an appropriate movement command. The corresponding movement commands are to move forward, turn left, or turn right, respectively. Before the queen moves in the appropriate direction, she transmits the movement command to the colonists. The colonists receive the signal, retransmit the signal to other colonists, and perform the corresponding movement.

The colonist controllers receive and transmit infrared movement signals from all four sides of their bodies. Because of this, the colony can be structured in a variety of patterns. Additionally, multiple receivers and transmitters add fault tolerance for the colony: If a controller cannot receive a signal in one direction for some reason, there is a good possibility that it will receive the signal through one of its other receivers. Figure 8.2 depicts a queen controller, Figure 8.3 a colonist controller, and Figure 8.4 the three possible walking modes.

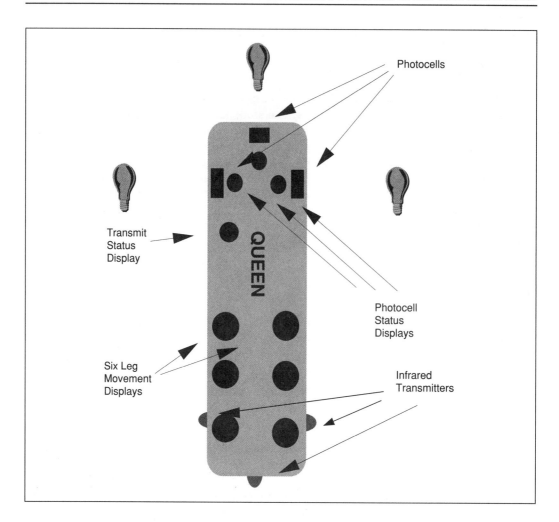

Figure 8.2. The queen Stiquito robot.

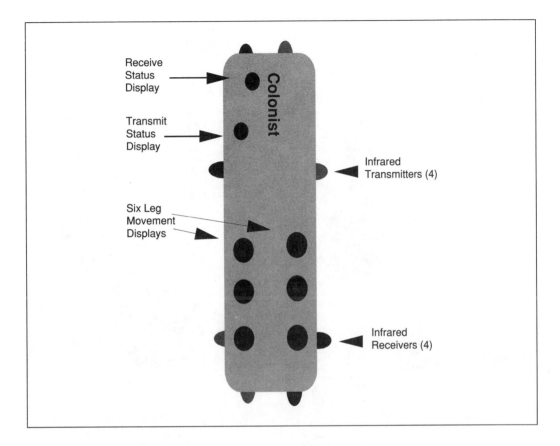

Figure 8.3. The colonist Stiquito robot.

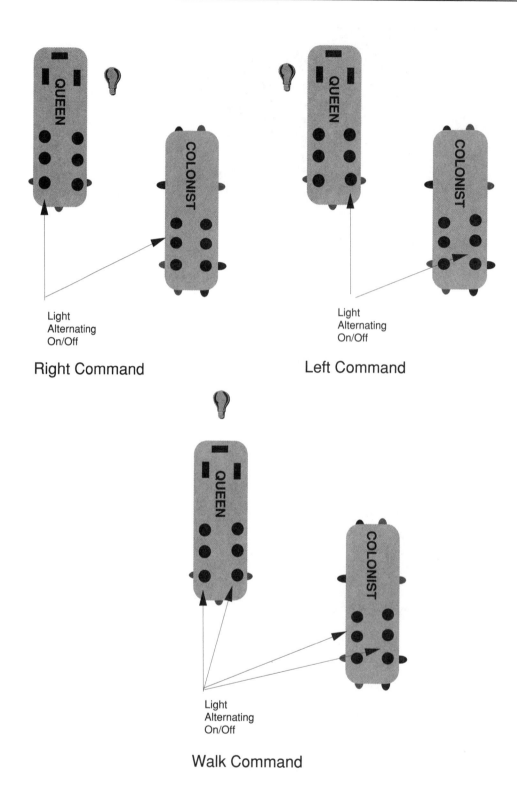

Figure 8.4. Modes of operation of the Stiquito robot colony.

Key Features and Functionality Improvements

Several functionality improvements had to be made to complete the project. Some of these include the following:

- *Creating a robotic colony structure.* All previous designs were built for one stand-alone robot.

- *Developing two levels of controller sophistication.* The queen controller is more sophisticated than the colonist controllers.

- *Generating transmit signals using an M68HC11 chip.* In previous designs, a transmitter chip was used to generate the appropriate transmit signals.

- *Creating controllers capable of two-way communication.* The previous micro-controlled Stiquito was capable of receiving infrared signals only, while the current controllers must receive and transmit these signals.

- *Allowing for an unlimited number of colonists in a colony.* This is possible because of the message-passing technique that was implemented in the software routines.

- *Freeing the colony from the user-operated handheld remote control.* In the current design, the queen controller generates movement commands by reading directional light levels.

- *Enhancing the overall design.* Some enhancements include mounting multiple transmitters and receivers on the controllers; adding circuitry to improve signal transmission and reception; and using LEDs to display the status of the transmission, the reception, and the photocell input process.

Hardware Development

The Motorola MC68HC811E2 Chip—The new controllers were developed using the MC68HC811E2 chip. This chip was chosen because of its functionality and because it was used in the previous microcontroller design.

The MC68HC811E2 is an 8-bit microcontroller chip containing five main ports and many peripheral capabilities[5]. Port A can be used for input-output (I/O) or for timer functions. Pin 7 of Port A can be used for pulse accumulator functions. Port B is an output port. All eight pins of Port B are used in the design. Port C can be configured as bidirectional I/O. Ports D and E were not used in the design. Features of these two ports include serial communications and analog-to-digital conversion. The microcontroller contains 2 kilobytes of on-board EEPROM (electrically erasable programmable read-only memory) and 256 bytes of random-access memory. Figure 8.5 shows a block diagram of the MC68HC811E2 microcontroller chip.

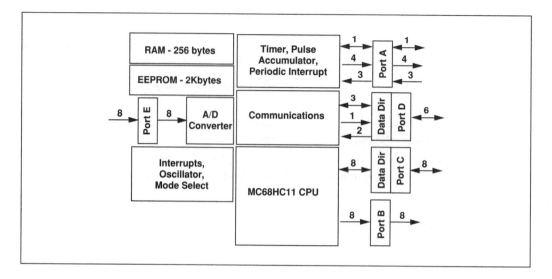

Figure 8.5. MC68HC811E2 block diagram.

Hardware Design—The MC68HC811E2 chip contains sufficient memory to hold the communication and control programs. The interrupt pins (XIRQ and IRQ) are disabled by connecting them to a 5-volt power source. The reset circuitry is controlled manually using an MC34064 chip. An 8-megahertz crystal is used for clock operation.

Two types of controllers were designed; they are shown in Figures 8.6 and 8.7. Figure 8.6 is the controller circuit for the queen. Port A was not used for this circuit. The first 6 pins of port B (PB0-PB5) are used to show movement commands via LEDs. PB6 is connected to an LED that is activated whenever the controller is in transmit mode. PB7 is used to control the MLED930 infrared-emitting diodes. The queen receives input from three MLD901 phototransistor detectors. These three detectors are connected to pins PC0-PC2.

Figure 8.7 shows the controller circuit for the colonist. In the circuit, the phototransistor detectors are removed and the receiver circuitry is added. The infrared signals are passed from the MR3150 infrared receivers to a 4049 CMOS (complementary metal-oxide semiconductor) inverter chip. The signal is then connected to pins 6 and 31 on the MC68HC811E2 microcontroller. Pin 6 is an input-capture (IC1) pin and pin 31 is PC0.

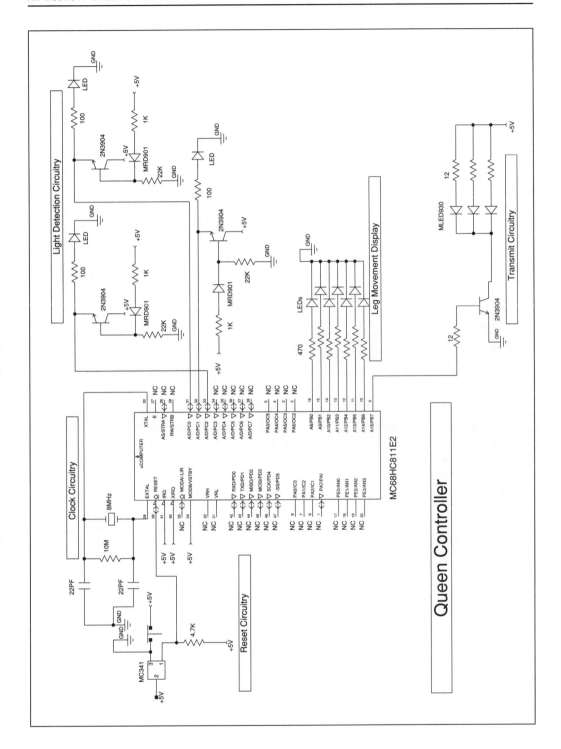

Figure 8.6. Schematic of the queen Stiquito robot.

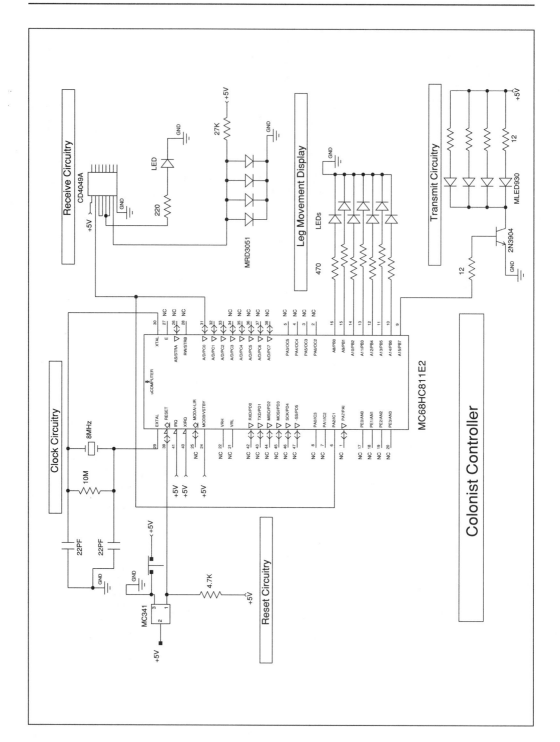

Figure 8.7. Schematic of the colonist Stiquito robot.

Software Development

Software Overview—Two separate software programs were written for this project. One controls the actions of the queen and the other controls the actions of the colonists. The code was written in M68HC11 assembly language and compiled using the AS11.EXE compiler. The software is divided into five primary routines.

1. **The main routine.** *Queen:* This routine polls to see which photocells are active. It then stores the information, jumps to the transmit routine, and then jumps to the appropriate leg movement routine. *Colonists:* The main routine for the colonists checks to see if a leg movement command has been received. Upon receipt of a valid leg movement command, it jumps to the transmit routine, then to the appropriate leg movement subroutine.

2. **The input-capture routine.** This routine is activated on the falling edge of a pulse through pin IC1. It then reads bit 0 of port C and decodes the incoming message.

3. **The timer-interrupt routine.** This routine is responsible for switching the LEDs on and off. In this project LEDs were used to simulate the legs of a robot.

4. **The transmit routine.** This routine is used to generate and transmit an infrared signal. In the original design this was accomplished through the use of an MC14497 transmitter chip.

5. **Leg movement routines.** Three separate subroutines are used to control the leg movements: forward pace, left turn, and right turn. Leg movements are demonstrated on the controller boards with LEDs.

The input-capture routine and timer-interrupt routine were not modified from the previous project's code. Refer to Nanjundan and Conrad[6] for details on these routines.

The Main Routine—The main routine for the queen polls pins PC0-2 to see if any are set high. These three ports are used in determining which movement command is to be transmitted. The following chart shows the pin/command relationships.

PIN NO.	DIRECTION	HEX VALUE
PC0	Forward	1A
PC1	Right turn	13
PC2	Left turn	19

Each of the three pins is checked. The main routine waits until one or more of the three pins are activated. If more than one pin is activated, priority is given in the following order: forward, left turn, right turn. Once the decision is made as to which command to send, the command is transmitted and the appropriate movement is shown on the queen.

The main routine for the colonists operates a little differently from that of the queen. The colonists wait for a movement command to be received. If a valid code is received, the colonists will retransmit that code and then act on it by performing the appropriate movements. Movement is shown in the LEDs that are connected to port B.

The Transmit Routine—The transmit routine is used by both the queen and the colonists. The transmit routine receives a hex value from the main routine and uses that value to generate a pulse code modulation (PCM) signal. This is the type of signal generated by the MC14497 transmitter chip that was used in the original design. An example of PCM coding is shown in Figure 8.8. PCM coding depends on which point of a bit is high or low. A 1-millisecond bit length is used; if the first half of the bit signal is high, it represents a logic 1. Otherwise, it represents a logic 0.

The first bit sent is the automatic gain control (AGC) bit. This bit will trip the interrupt in the input-capture routine. Next, two bit times of 2 milliseconds are sent with no signal. Then a logic 1 is sent to represent the start bit. Finally, the hex code that has been previously stored in the program-variable PCM is sent. Only the first six bits of the hex code are transmitted. The lowest bit is transmitted first.

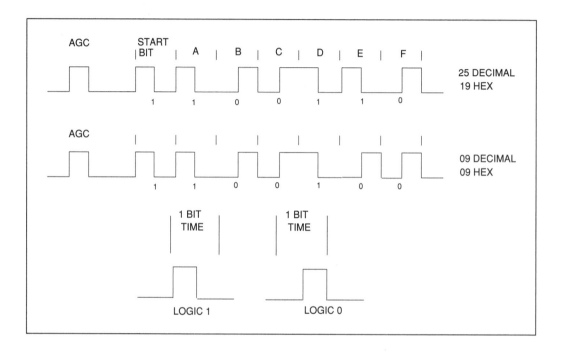

Figure 8.8. Pulse code modulation.

The Input-Capture Routine—This routine was written by the original programmer. It uses two pins on the M68HC11 to acknowledge and interpret the infrared signal being received. The first pin, IC1, is triggered on the falling edge of an AGC pulse. This pulse alerts the microcontroller to start watching for values on pin PC0. Approximately 2 milliseconds after the AGC pulse is read, PC0 reads as a logic 1. This is the start bit. A small delay is used to position the read of the next six bits. The start bit is discarded. PC0 is read six times, once every millisecond, to capture the hex code that was sent. As each bit is read, it is rotated into a register. When all six bits have been read, the code is stored to a variable that represents the appropriate movement command.

EEPROM Programming on an MC68HC811E2 Chip—The MC68HC811E2 microcontroller chip was programmed using an M68HC11 Evaluation Module (EVM). The EVM can program either plastic leaded chip carrier (PLCC) or dual in-line pin (DIP) microcontrollers. This design uses the DIP version. A personal computer (PC) is also necessary to program the chips. The PC communicates with the EVM through a serial port. The software communications program Kermit was used to establish the connection between the PC and EVM and to download programs to the chips. The LOAD command loads an assembled program (.S19 file) into memory on the EVM. The PROG command then transfers the program to the DIP microcontroller.

Problems Encountered During Project Development

It is not unusual for problems to arise during a major project. The following is a summary of some of the problems that were encountered during this project.

- There were many small hardware-related problems. This was primarily due to the limited hardware experience of the two team members. Fortunately, these types of problems rapidly decreased as the students' knowledge of hardware grew.

- Initially, the infrared signals that were being received were very weak. Consequently, the controller was extremely sensitive. The infrared transmitters and receivers had to be placed right next to each other for the receiver to acknowledge any signal. Even then the signal was still too weak to be used to generate a correct response. Several actions were taken to solve this problem, including changing to a better-quality infrared receiver, adding a transistor to the transmit circuitry to boost the transmitted signal, and adding components such as capacitors and inverters to the receive circuitry to clean up the received signal.

CONCLUSIONS AND FUTURE WORK

This project was chosen to accomplish three primary goals: (1) To further the current research in the area of robotic communications; (2) to further Dr. Conrad's research on autonomous robots; and (3) to provide project team members an opportunity to gain hardware skills and experience.

Completion of this project brought the accomplishment of all three goals. A communications system for autonomous robots was designed. A colony structure, consisting of a queen and unlimited colonists, was designed from the previous single-controller design. Two-way communication within the colony was established through the use of infrared signals.

In addition to the basic communication system, the queen was designed to receive input from three mounted phototransistor detectors. This enhancement frees the colony from the infrared remote control that was used in the previous stand-alone microcontroller design.

The project participants are undergraduate computer system engineering majors, and this project gave them an opportunity to learn, hands-on, about hardware design. There was a very high learning curve associated with this project; the participants soon learned that circuit theory is not quite the same as real life. A couple of chips were destroyed and a few transistors became overheated, but, in the end, the participants came away with a much greater command of hardware design. Much was also learned about infrared signaling, pulse code modulation, and M68HC11 circuitry and assembly code.

Future design possibilities for this project include the following:

- Increasing the maximum distance allowable between controllers. There are many ways that this might be accomplished. One way would be to redesign the transmitter circuitry to put out a stronger infrared signal. Another possibility would be to use a filter to reduce radio frequency interference.

- Researching new communication techniques, such as using a tower repeater. Tower repeaters that can receive and boost signals are available on the market. Current uses for tower repeaters include improving the distance of normal television and VCR remote controllers.

- Tracking or polling active controllers.

- Adding more possible light commands. This could be useful for expansion projects involving new robot designs.

- Adding an octagon decoder. Currently, six bits of the M68HC11 chip's B port are used to control the six legs. More-sophisticated robots will probably require more output lines. An octagon decoder will help meet future expansion needs.

- Widening the transmission arc. The current arc of the infrared signal is very narrow. An enhancement might include widening the arc.

- Increasing the transmission time. The length of each transmitted bit is currently 1 millisecond. This was the value used in the original microcontroller design. In that design the value was determined by the MC14497 chip. In the current design the signal length is generated by the software routines. Therefore, the length of each bit can be modified. Increasing the signal transmission time would probably decrease the chance of errors at greater distances.

ACKNOWLEDGMENT

The authors would like to thank Motorola Corporation for its kind donation of semiconductor parts and development boards.

REFERENCES

1. Robots work better when they communicate. 1993. *IEEE Expert: Intelligent Systems and Their Applications*, Vol. 8, No. 5: 77.

2. Mills, Jonathan W. 1992. Stiquito: A small, simple, inexpensive hexapod robot. Computer Science Dept., Indiana Univ., Technical Report No. 363a. See also Chapter 2.

3. Nanjundan, Mohan and James M. Conrad. 1993. A M68HC11 microcontroller-based Stiquito controller. Computer Systems Engineering Dept., Univ. of Arkansas, Technical Report 1993-1. See also Chapter 7.

4. Ciarcia, S. 1987. Build an infrared remote controller. *Byte*, Feb.: 101–110.

5. Motorola, Inc. 1991. *M68HC11 reference manual.* Englewood Cliffs, N.J.: Prentice Hall.

6. Nanjundan and Conrad. A M68HC11 microcontroller-based Stiquito controller.

Chapter 9

A General-Purpose Controller for Stiquito

Shyamsundar Pullela

This chapter describes a general-purpose digital controller for the Stiquito robot. The controller can steer the direction of walking, synchronize the leg movements, and be programmed for different gaits. It can also be given external clocking signals to control the legs.

INTRODUCTION

This chapter describes a general-purpose digital controller that can be programmed to produce different gaits in the Stiquito robot. This controller was developed primarily so that researchers could concentrate on developing different gaits for Stiquito rather than worrying about how to implement them.

Implementing different gaits requires the following:

- Some way of activating the legs in varying order so that different gaits can be tested.

- Some way of controlling the displacement of the legs. Displacement is used to turn the Stiquito in a particular direction. For example, less displacement in the left legs and more displacement in the right legs makes the Stiquito turn left.

- Some way of synchronizing the legs of Stiquito with an external signal so that multiple Stiquitos can work together on a single task, signaling each other of their status.

The controller described in this chapter has all these capabilities and the option of using either an internal clock or an externally provided clock. The controller was designed using an Actel FPGA (field-programmable gate array) chip. The functionality of the controller is explained in subsequent sections.

SINGLE-LEG CONTROLLER

The controller for a single leg is very simple. The logic for a single leg is shown in Figure 9.1.

The state machine of the controller will loop in the start state as long as the leg is inhibited. When inhibit is FALSE, the displacement is loaded into the counter and is counted down. When the count reaches zero the machine goes into a synchronizing state, where it waits for the SYNC signal. After synchronization it goes back to the initial state.

Note that when inhibit becomes FALSE the value present in the displacement lines will be taken as the displacement. Any external circuit that controls this leg should ensure that the correct displacement value is present in the lines when the leg is activated.

The schematic for a single-leg controller is shown in Figure 9.2.

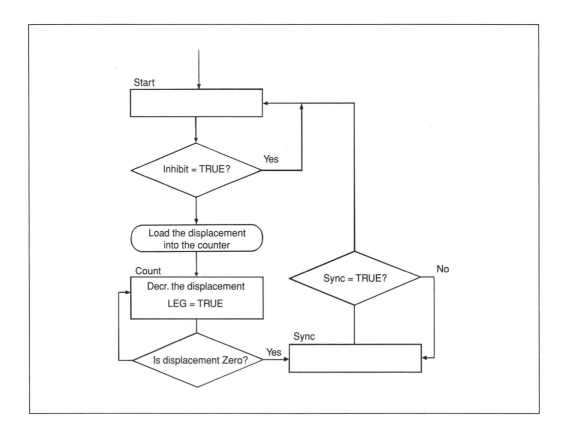

Figure 9.1. Logic for movement of a single leg.

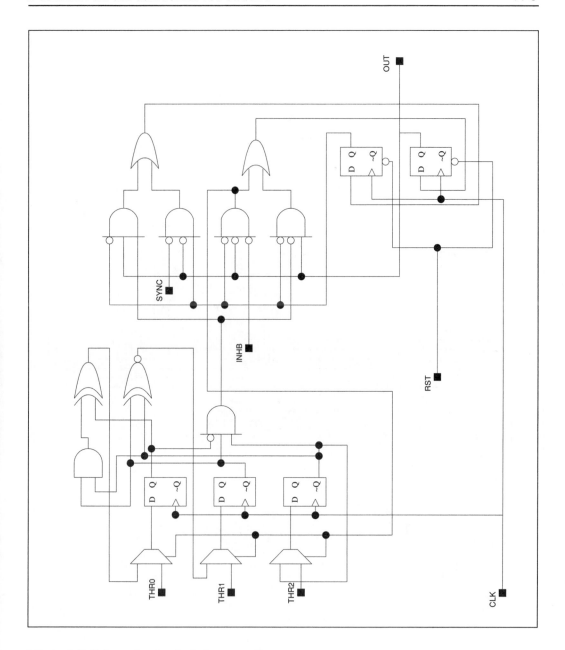

Figure 9.2. Schematic of a single-leg controller.

CONTROLLER FOR SIX LEGS

The controller for six legs is nothing more than a collection of six single-leg controllers, except that the displacement values of all the left legs are connected and the displacement values of all the right legs are connected. All six single-leg controllers run with the same clock and their reset signals are connected.

The schematic for the six-leg controller is shown in Figure 9.3.

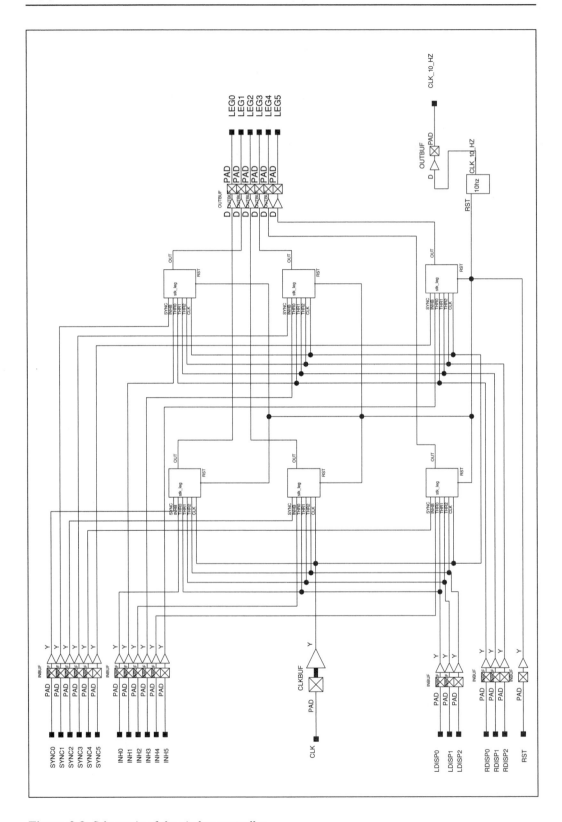

Figure 9.3. Schematic of the six-leg controller.

THE INHIBIT SIGNALS

Inhibit signals are used to control the activity of the Stiquito legs. As long as the inhibit signal for a leg is TRUE, that particular leg is inactivated; thus, the signal on the leg's output pin would be FALSE.

When the inhibit signal of a leg becomes FALSE, the controller takes the displacement value present in the lines and asserts TRUE for the leg's output signal, based on the number of clock pulses that had been latched when the inhibit signal went FALSE. The controller then enters a synchronizing state, where it waits for a SYNC signal.

THE SYNC SIGNALS

After maintaining the TRUE value on the output of the leg for the displacement number of clock cycles, the state machine enters a synchronizing state. It waits in this state until it receives a synchronizing signal and then returns to the initial state. This synchronizing signal is not necessary in an ideal setup; in practice, however, all the legs do not have the same mechanical properties. To overcome this problem, a synchronizing signal is required so that all the legs enter the initial state together.

The SYNC signals were introduced into the design to synchronize one Stiquito with another. When Stiquitos are exhibiting cooperative behavior it is essential that they synchronize with one another; these signals can used for that purpose.

THE DISPLACEMENT SIGNALS

The Stiquito controller has two sets of signals for controlling the displacement of the left and right legs. Displacement is a measure of how many clock cycles (how long) the leg should be activated. A larger displacement will deliver current to the leg for a longer time, which will contract the Nitinol wire more and cause the leg to swing back farther.

For the sake of implementing the controller in the digital domain, the maximum displacement is divided into eight equal parts, and displacement is represented by an integer between 0 and 7. Maximum displacement, 7, will activate the Nitinol wire for about 1 second, and a displacement of 0 will not move the leg.

Displacement is used to control the directional movement of Stiquito. For example, if a displacement of 2 is given to the left legs and a displacement of 7 to the right legs, the right legs will move more than the left legs, and Stiquito turns left. Likewise, Stiquito turns right if the values of the left and right displacements are switched.

The displacements are used only for turning the Stiquito. In designing the controller, a situation could not be imagined in which the legs on one side required different displacements; therefore, the displacements of all the legs on each side are connected.

THE LEG-CNTR SIGNALS

There are six LEG-CNTR signals, one for each leg. The legs are numbered from top to bottom, left to right, from 0 to 5. All the left legs are even numbered, the right ones odd. The current output of the Actel pins is not sufficient to drive the legs effectively, so the LEG-CNTR signals are driven through transistors to ensure sufficient current. The LEG-CNTR pin will have a TRUE value when the leg corresponding to that pin is supposed to be activated. While the pin has a TRUE value, current is passed through the Nitinol wire connected to the corresponding leg.

THE CLK-10-HZ PIN AND THE CLK-IN PIN

A 10 Hz clock is provided on the chip for convenience. The clock is generated by a loop of an odd number of NOT gates. It is then frequency-divided by a sequence of T flip-flops. The CLK-IN pin signal is connected to the clock that is to control the state machine of the controller. The machine can be connected to the CLK-10-Hz pin or to an external clock.

AN EXAMPLE: THE TRIPOD GAIT

The following describes how a tripod gait can be obtained using the controller. Other gaits can be obtained in a similar way.

A tripod gait is achieved when the legs are divided into two sets, each set containing two legs on one side and the middle leg of the other side. For example, in this controller legs 0, 3, and 4 constitute one set and legs 1, 2, and 5 make up the other. For purposes of illustration, consider a displacement of 1 for each leg. The timing diagram is shown in Figure 9.4.

CONCLUSION

A programmable digital controller was designed for Stiquito. The controller can be programmed to create different gaits and different speeds in the robot. Its main purpose is to give researchers a simple and convenient controller so they can concentrate on the actual software control of gaits rather than hardware circuitry and control.

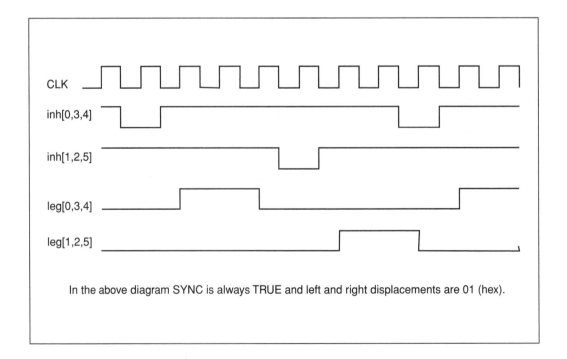

In the above diagram SYNC is always TRUE and left and right displacements are 01 (hex).

Figure 9.4. Timing trace of control circuit with a tripod gait.

Chapter 10

SCORPIO: Hardware Design

John K. Estell, Timothy A. Muszynski, Thomas A. Owen,
Steven R. Snodgrass, Craig A. Szczublewski,
and Jason A. Thomas

The goal of the SCORPIO robotics project is to develop a microcontroller system for the Stiquito robot for the purpose of performing independent, intelligent operations. This task requires the collaborative efforts of both hardware and software designers. This chapter covers the hardware aspects of the SCORPIO design, which have been successfully implemented and tested. The system is based on the 80C32 microcontroller and contains external random-access memory (RAM) and read-only memory (ROM). Interfaces are provided for driving the Stiquito leg circuitry, using sensors, and performing communications. From this work SCORPIO has grown into a viable testbed for experimentation in sensor design, robotics, and artificial intelligence (AI).

INTRODUCTION

As originally designed, the Stiquito robot can carry up to 50 grams of weight and travel at a speed of approximately 3 to 10 centimeters per minute.[1] The goal of this project was to develop a tetherless Stiquito unit capable of performing intelligent operations. To implement this, a microcontroller system was added to the original Stiquito design. This microcontroller system was developed by students enrolled in the Microcomputer Systems II course offered at the University of Toledo in the spring 1993 quarter. Named SCORPIO (an acronym that stands for Stiquito Controller using On-board Resources to Provide for Independent Operation), the system was produced by the students using the codesign process.[2] Two research groups were formed within the class: one for hardware, and one for software. Each research group was responsible for providing proper communication to the other group as to the requirements and capabilities of the emerging properties that resulted from the concurrent hardware and software design processes. This chapter focuses on the development of the hardware portion of the design.

CONTROLLER DESIGN

In developing the control circuitry of the system, the first step was to select a micro-controller. The 8051 family of microcontrollers was investigated because of the variety of features available within the family, market popularity, and cost-effectiveness.[3,4] The next step was to select an individual controller from this broad family. The 80C32 was eventu-ally chosen because it provides 256 bytes of internal RAM as well as the ability to control up to 64 kilobytes of both program memory and data memory in a Harvard architecture.[5,6] This provided more than enough flexibility for the project design. Once the microcon-troller was selected, a logic family had to be chosen. As this project is battery-powered, power consumption was a significant issue that had to be addressed. By selecting the 74HC high-speed CMOS (complementary metal-oxide semiconductor) logic family, the power consumption of the controller circuitry was minimized.

The 8051 family of microcontrollers uses a Harvard architecture, which is not particu-larly useful in this application, as it is not conducive to program modification by the end user (the program segment is read-only). If the 80C32 microcontroller is used as designed, the erasable programmable read-only memory (EPROM) on the board containing the pro-gram has to be removed from the board and reprogrammed for every change to the code that the user wants to make. Since one of the goals of the project was to design a flexible, yet inexpensive, robotic platform, this memory limitation was not acceptable. The solu-tion, shown in Figure 10.1, was to provide a section of the data memory that overlapped the program memory. This was done by ANDing the PSEN and the receive-data (RD) lines together to provide the read signal for the static random-access memory (SRAM). The SRAM was then set up to occupy the higher 32 kilobytes of both program and data mem-ory, with line 7 of port 2 (P2.7) used as the select line and the remaining 15 lines of ports 0 and 2 used as the address bus. This approach provides the SCORPIO design with 32 kilo-bytes of SRAM and 32 kilobytes of EPROM, resulting in 64 kilobytes of address space. For each component of the memory system one 32-kilobyte-by-8-bit memory integrated circuit (IC) was selected to provide a flexible memory system with a minimal chip count.

Since the system now has a user-programmable segment of program memory, the hard-ware must provide an easy method of programming this memory. This requires some means of communication with the outside world. Fortunately, the 80C32 has a built-in universal synchronous-asynchronous receiver/transmitter (USART). This USART was connected with a minimal amount of external analog circuitry for a combination infrared/direct-connect modem. The circuitry can either broadcast the signals via an infrared light-emitting diode (IR LED) or through a wire. This gives the system the flexibility of either tethered or line-of-sight communication. Additionally, the IR modem can employ line-of-sight commu-nications with either a host computer or another robot. This provides SCORPIO with a design that is sufficiently flexible for AI and emergent behavior experimentation.

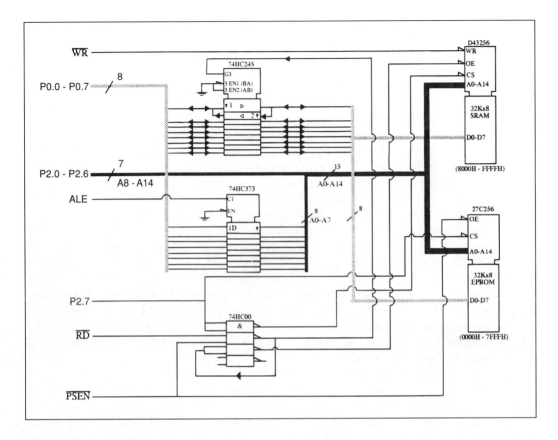

Figure 10.1. SCORPIO memory interface.

DESIGN CONSTRAINTS

In order to retain the major properties of the original Stiquito, such as the nitinol actuator wires, some design constraints had to be addressed. The original chassis design and leg assembly allowed for only 50 grams of payload weight. One of the simplest ways to handle this problem is to add more legs to increase the loading capacity. This approach, however, is not without side-effects. The nitinol wire used in the legs is 100 microns in diameter and has a recommended power rating of 4.86 watts per meter.[7] In the original Stiquito design, the effective wire lengths were approximately 40 millimeters. Therefore, each nitinol actuator wire required 194 milliwatts to contract, resulting in a draw of 180 milliamps from the power supply passing through the wire. To support the estimated weight of the controller, a 12-leg system would be needed. Assuming a worst-case duty cycle of 33 percent and a maximum of half of the legs active at any one time, the peak current that would be required is about 1 amp, with an average current for continuous walking of 360 milliamps. Additionally, the controller circuitry would also require current on the order of 100 milliamps. Thus, a conservative power supply design would have to be able to continually source about 500 milliamps. To accommodate the increased demand, a larger battery-based power supply would be necessary, which in turn would increase the overall weight of the robot. In order to resolve the weight-versus-current-demand problem, modifications to the leg and chassis systems were deemed necessary. The solution was to develop a reinforced, pin-jointed leg assembly.[8] This design solved the problem by significantly increasing the loading capacity of each leg. With the development of the new leg and chassis, an eight-leg design was chosen for SCORPIO. The robot can now carry its required load without placing unreasonable demands on the power supply.

Because SCORPIO uses batteries, it was desirable for the power supply to be rechargeable to reduce the long-term cost of operating the robot. The most commonly available type of rechargeable battery is nickel-cadmium (NiCd). NiCd batteries have high current capability and are lightweight compared to their alkaline and carbon-zinc counterparts.[9] For these reasons, NiCd was the obvious choice, but the choice of what cell size or sizes to use was more difficult. Unless a step-up DC-to-DC (direct current to direct current) converter were used (something that was initially unavailable to the research group), a minimum cell voltage of 4.5 volts would be required to power the processor. A voltage regulator would be nearly mandatory in this situation, but it adds another constraint. Every linear voltage regulator has a dropout voltage, defined as the minimum difference between the input voltage and output voltage required to maintain regulation. Therefore, to allow a reasonable voltage margin for the 5-volt regulator, a 9-volt NiCd cell was chosen to power the control circuitry. Battery life for the ICs depends on the dropout voltage of the regulator as well as the current drawn by the ICs, since the battery voltage will slowly decrease as it discharges. Using a low-dropout regulator would extend the battery life; +5-volt regulators are available with dropout voltages as low as 150 millivolts (for example, the MAX667). Since there were no low-dropout regulators available to the group at the time of this design, the LM317 was used.

A dual-voltage power supply was designed for SCORPIO, as it was determined that powering both the controller circuitry and the leg circuitry from a single 9-volt cell was inefficient. Using a dual power supply has two advantages: It prolongs the life of the 9-volt cell, and it eliminates a second voltage regulator from the system that the leg circuitry would otherwise require. A NiCd AA cell has a voltage of 1.2 volts and a 2-amp current rating; placing two in series gives an acceptable voltage for the leg circuitry of 2.4 volts and weighs only 48 grams.[10] The 9-volt cell weighs about 40 grams, yielding a total power source weight of 88 grams. This design keeps the components required for the power system to a minimum, provides SCORPIO with an operational life of approximately 1 hour, and makes the system completely rechargeable. In addition, both of the cell types used are commonly available, making cell replacement simple.

Figure 10.2 shows the design of the leg actuator circuitry, which incorporates the 2.4-volt source. The legs are controlled by a latching system that can handle up to 40 individual legs, organized into five banks of eight legs each. Each leg bank consists of an octal latch. The current SCORPIO design uses only one of the banks; the remaining banks are available for future expansion, such as more legs or more degrees of freedom per leg. (The second latch shown in Figure 10.2 is intended for providing a second degree of freedom to the current leg system.) The legs are accessed via memory-mapped input/output locations in the lower half of the external data memory. In order to save weight and circuitry, the addresses are not fully decoded. The data inputs of the latch are connected to the data bus; each data output of the latch is used to control one leg circuit. With this method, up to eight legs can be activated at one time.

SENSOR DESIGN

The SCORPIO sensory system has the capability of handling up to eight devices. The system has two major parts: the sensory control logic and the sensors. Shown in Figure 10.2,

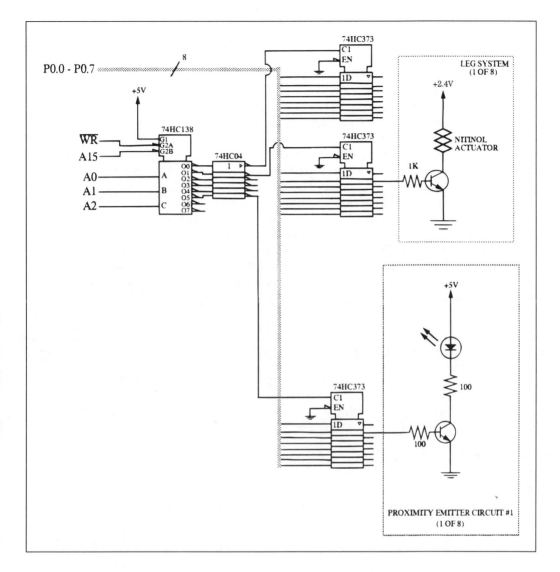

Figure 10.2. SCORPIO analog output circuitry.

the sensory logic consists of a 3-to-8 decoder, an octal latch, and an inverter. The 3-to-8 decoder determines whether the program is trying to access the sensors. The inverter takes the output of the 3-to-8 decoder and inverts it for use as the strobe signal for the latch. The latch stores the value representing the pattern of sensors that the user program wishes to activate. On a logic 1, the latch forces the NPN transistor into saturation, allowing current to flow through the LED, thus activating it. On a logic 0, the NPN transistor is in cutoff mode, preventing the flow of current to activate the LED. In this manner the user program is able to activate any or all of the sensors. This is useful especially if the user wishes to implement a custom sensor sweep pattern.

For this particular implementation, a proximity detector system was designed. Four sensors are used. Two are oriented to point forward, located on the side edges of the board. The other two are also located on the side, oriented perpendicular to the direction of travel, with one looking to the left and one looking to the right. The actual proximity sensors are designed to sense objects up to 30 centimeters away, and to be resistant to changes in the ambient light level of the room. Each sensor contains two major components: an IR emitter circuit and an IR detector and comparator circuit. The emitter, shown in Figure 10.2, consists of a transistor and an IR LED. It is controlled by an active high signal from the sensory control logic circuitry. The IR detector circuitry shown in Figure 10.3, consists of two IR photodiodes, an operational amplifier, and a comparator. The negative input of the comparator is connected to the detector circuit that senses IR light coming from the emitter. The positive input of the comparator is connected to a detector circuit that senses the ambient light level in the room. This detector is adjusted to be at a slightly higher voltage than the other detector when the IR emitter is off. When the emitter is turned on, the IR sensing detector will increase in voltage relative to the intensity of the reflected beam, while the ambient detector's voltage will remain constant. If an object is close, the voltage will rise sufficiently to change the output of the comparator, thus "sensing" the object. The outputs of the four comparators in this design are connected directly to port 1 of the 80C32 to provide easy access to each sensor's status and to reduce the number of components. The comparator's outputs could also have been multiplexed to minimize the number of port lines used.

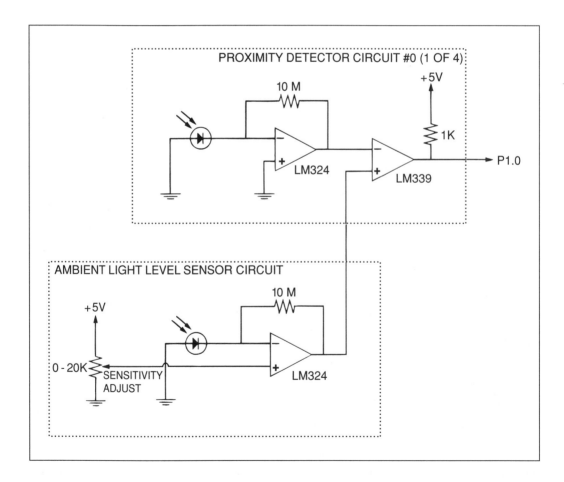

Figure 10.3. SCORPIO IR proximity detector circuit.

The finished SCORPIO robot is shown in Figure 10.4.

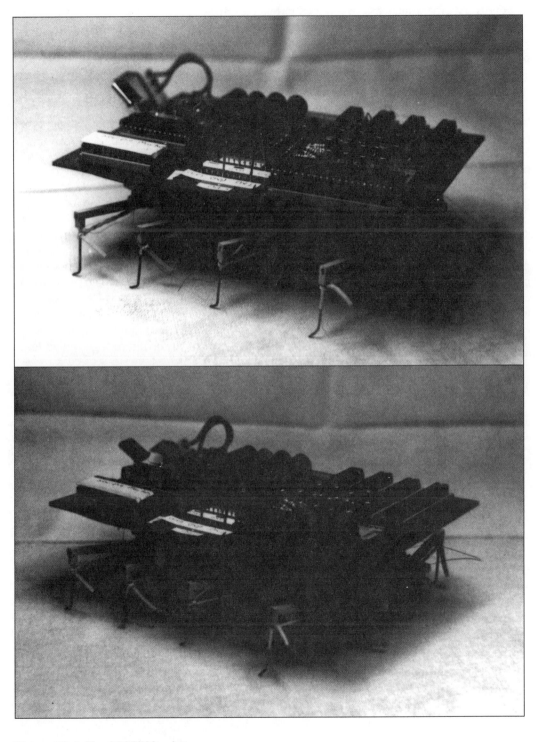

Figure 10.4. The SCORPIO robot.

SUMMARY AND CLOSING COMMENTS

SCORPIO has shown itself to be a viable robotic system. Its development has presented an opportunity for students to become experienced with some of the leading hardware production tools, such as printed circuit board auto-routers and in-circuit microcontroller emulators. The availability of these and other tools made it possible to develop a working prototype system before any construction began on a final dual-sided printed circuit board. The prototype offered flexibility to change parameters of the project before the design was made final. The circuitry for SCORPIO has been successfully tested and is now operational. The robot has a top speed of 6.67 centimeters per minute, and a steady speed of 6 centimeters per minute; this compares favorably with the original Stiquito robot's speed of 6.35 centimeters per minute.

Several improvements are envisioned for the SCORPIO hardware design. The large power draw of each leg movement quickly depletes typical NiCd batteries. Therefore, to facilitate long-term operation, an on-board battery charger using solar cells would be a beneficial addition. The system would recharge itself upon detecting a low-battery condition, then proceed with its assigned tasks. Although not as important to general operation, a leg system that can move the leg vertically as well as horizontally would allow for travel on rougher terrains than is currently possible. The current version of the proximity detector is primitive in its design. Plans are under way to use an ultrasonic ranging unit like that used in a Polaroid camera. The unit would be mounted in a device providing two degrees of freedom, which would allow SCORPIO to "see" its environment. Finally, DC-to-DC converters would provide a large advantage over voltage regulators when attempting to maintain operation of the design as battery strength decreases. Samples of various converters have been obtained and future designs will incorporate a DC-to-DC converter for voltage regulation.

REFERENCES

1. See Chapter 2.

2. Estell, J.K., and T.A. Owen. 1995. Experiencing the codesign process: Microcomputer Systems II laboratory. *SIGCSE Bulletin*, Vol. 27, No. 1: 34–38.

3. MacKenzie, I. 1992. *The 8051 microcontroller.* New York, N.Y.: Merrill.

4. Sperry, T. 1993. Tools for 8051 development. *Embedded Systems Programming*, Vol. 6, No. 3: 49–60.

5. Intel Corporation. 1992. *Embedded microcontrollers and processors, vol. 1.* Mt. Prospect, Ill.: Intel Corporation.

6. Siemens Components Inc. 1990. *8-bit single-chip microcontroller handbook.* Santa Clara, Calif.: Siemens Components Inc., Integrated Circuits Division.

7. Gilbertson, R. 1992. *Motorless motion! Working with shape memory wires*, 2d ed. San Anselmo, Calif.: Mondo-tronics, Inc.

8. See Chapter 4.

9. Radio Shack. 1990. *Enercell battery guidebook*, 2d ed. Richardson, Tex.: Master Publishing, Inc.

10. Ibid.

Chapter 11

SCORPIO: Software Design

John K. Estell, Christopher A. Baumgartner, and Quan D. Luong

The goal of the SCORPIO robotics project is to develop a microcontroller system for the Stiquito robot for the purpose of performing independent, intelligent operations. This task requires the collaborative efforts of both hardware and software designers. This chapter covers the software aspects of the SCORPIO design. The system is based on the 80C32 microcontroller and contains external random-access memory (RAM) and read-only memory (ROM). A control language was developed to program the robot, using an interpreter residing on board SCORPIO. The interpreter is part of a kernel that controls all aspects of the robot's operation. From this work, SCORPIO has grown into a viable testbed for experimentation in sensor design, robotics, and artificial intelligence (AI).

INTRODUCTION

The Stiquito robot[1] was the inspiration behind the development of a robot operated by an on-board microcontroller system, designed by students enrolled in the Microcomputer Systems II course offered at the University of Toledo in the spring 1993 quarter. Named SCORPIO (an acronym for Stiquito Controller using On-board Resources to Provide for Independent Operation), the system was produced using the codesign process.[2] Two research groups were formed: one for hardware, and one for software. Each research group was responsible for providing proper communication to the other group as to the requirements and capabilities of the emerging properties that resulted from the concurrent hardware and software design processes. This chapter focuses on the development of the software portion of the design.

The software design for SCORPIO was open-ended; the resulting product required as much creativity as logic to produce an efficient, flexible, and expandable software foundation for the developed SCORPIO hardware.[3] The initial stages of design saw two major needs. First was the ability to write SCORPIO control programs with relative ease. The SCORPIO control language (SCL) was developed to allow programmers to write control programs without having to know all of the technical details involved in controlling the robot. Second, programs were needed for the operation of SCORPIO. A kernel was developed for SCORPIO's on-board microcontroller to provide an interpreter for the execution of SCL programs, a monitor, and a communications package based on the XMODEM protocol.

SCORPIO CONTROL LANGUAGE

Many things were considered when developing SCL. First was the need for simplicity in the construct of the language. With this in mind, a language was developed with few commands and only two addressing modes. Another goal was to have complete control over SCORPIO's movement with the language. Finally, it was desired that the language possess sufficient robustness such that elaborate programs could be written. It was decided that the language would be implemented using an interpreter residing on board SCORPIO that would operate via code generated by an assembler. This approach allows SCL to be independent of the hardware used to control the robot. Consequently, if a hardware change is made, only the interpreter and subsequent assembly language routines will need modification.

The current version of SCL is similar to assembly language. The elements of the language, shown in Table 11.1, are mnemonic and are usually followed by a single operand. The reason behind this is that it is much easier to write an assembler for an assembly-like language

command	description
BCH	branch to location
CMP	compare the accumulator with the value
DEC	decrement
FWD	move forward
INC	increment
JEQ	jump if equal to
JGE	jump if greater than or equal to
JGT	jump if greater than
JLE	jump if less than or equal to
JLT	jump if less than
JMP	unconditional jump
JNE	jump if not equal to
JSR	jump to a machine language subroutine
LDA	load the accumulator
LFT	turn left
NOP	no operation
RET	return to monitor control
RGT	turn right
STA	store the accumulator
STS	set step size
TST	test the accumulator with the value

Table 11.1. SCORPIO SCL elements.

than it is to write a compiler for a structured language. For this particular design, only a minimal set of language elements was implemented. SCORPIO's movements are controlled by commands for going forward and turning left or right. Decision-making ability is available from the jump, test, and compare instructions. Simple arithmetic operations are also available. Furthermore, the language allows the user to call subroutines written in machine language from a user program. The user-supplied routine can be assembled using any 8051 assembler and then loaded into memory using the SCORPIO communications package. This feature was added for testing and debugging new SCORPIO functions and for implementing custom-designed hardware drivers. Plans are currently being made to expand SCL to include "intelligent" or "self-awareness" commands, such as the need for battery recharging and performing proximity detection using infrared (IR) sensors.

The four movement commands all require a single 1-byte operand. This operand is either specified using immediate mode addressing or assumed to be the current value of the accumulator. This byte, along with the current step-size, determines how many steps are to be taken. The number of steps is the product of the two bytes; therefore, SCORPIO has the ability to take up to 65,025 steps with a single movement command. A SCORPIO "step" consists of two physical steps, each of which moves four legs in a pattern similar to that of Stiquito's tripod gait.[4] This movement scheme allows SCORPIO to cover large or small distances with a single command.

A sample SCL program is given in Figure 11.1. The program begins by having SCORPIO take 400 steps forward and then turn to the right. At this point the program enters a loop whose first instruction is to clear the accumulator. Next, it jumps to a user-supplied 8051 subroutine at the address CAB0h and, upon returning, checks to see if the accumulator is still zero. If the user-supplied subroutine has changed the value in the accumulator, then SCORPIO will use this value to determine how far to proceed forward; otherwise, the program exits and control of SCORPIO is returned to the monitor.

Programs written in SCL are parsed by the SCORPIO language assembler (SLA). Written in C language for portability, the assembler breaks down the mnemonics and

```
START:   STS      #0A      ; set step-size to 10
         FWD      #28      ; forward 40 steps
         RGT      #0F      ; turn right 15 steps
LOOP:    LDA      #00      ; clear the accumulator
         JSR      &CAB0    ; jump to 8051 subroutine at CAB0h
         CMP      #00      ; compare the accumulator to zero
         JEQ      STOP     ; stop if it is equal to zero
         FWD      A        ; forward # of steps in accumulator
         BCH      LOOP     ; loop again
STOP:    RET               ; return to monitor
```

Figure 11.1. Example of an SCL program.

operands into 8-bit codes. The difference is that, in other assemblers, the values represent the machine code for a particular processor, while the codes produced by the SLA represent routines and data to be handled by the interpreter. Figure 11.2 shows the listing file generated by the assembler from the code given in Figure 11.1. The first column in the listing file contains the address of the current instruction. All jump instructions, however, except for JMP, are relative to the current location; therefore, the code can be placed anywhere in memory if this instruction is not used. To the right of the address is the hexadecimal opcode(s) of the instruction; these opcodes are also stored in an object file. Finally, an address table is printed listing the address of each label present in the program.

```
                    .start
0000 31 0A    START: STS   #0A      ;set step-size to 10
0002 01 28           FWD   #28      ;forward 40 steps
0004 11 0F           RGT   #0F      ;turn right 15 steps
0006 97 00    LOOP:  LDA   #00      ;clear the accumulator
0008 48 CA B0        JSR   &CAB0    ;jump to 8051 subroutine at CAB0h
000B A1 00           CMP   #00      ;compare the accumulator to zero
000D 71 04           JEQ   STOP     ;stop if it is equal to zero
000F 00              FWD   A        ;forward # of steps in accumulator
0010 42 F5           BCH   LOOP     ;loop again
0012 40       STOP:  RET            ;return to monitor
                    .end


   -- Labels --


START         - 0000
LOOP          - 0006
STOP          - 0012
```

Figure 11.2. Example of an SLA listing file.

SCORPIO KERNEL

The SCORPIO kernel is composed of three items: the interpreter, the monitor, and the communications package. The SCORPIO interpreter is a program that reads SCL opcodes from external RAM and executes the corresponding machine language procedures stored in ROM. The interpreter parses each instruction before executing the appropriate routine; therefore, a program written in SCL will not be as fast as one written entirely in the machine language of the microprocessor. The interpreter should be as efficient as possible so the performance of SCORPIO is not adversely affected. Because of this, the interpreter was written entirely in 8051 assembly language instead of a high-level language. This optimizes the operation of the interpreter, resulting in increased speed and efficiency.

The first step in the design of the interpreter was to develop pseudocode and a flowchart for the parsing of the instruction byte. This illustrated how the interpreter would actually work. Next, several small 8051 assembly language programs were written to test the various instructions for comparisons and jumps. These programs were executed on a software version of an 8051 emulator. The actual interpreter was then written using top-down methodology. The "skeleton" was written first so that the instruction was parsed; however, the actual implementation of each instruction was not important at that time. It was made more efficient by taking advantage of the Boolean processing facilities available with the 8051 microcontroller family. This allows one to perform logic functions on single bits as well as have up to 210 bit-addressable locations within the 8051's internal RAM.[5] Single-bit comparisons are used in parsing the instruction to minimize the amount of code needed to test the value of the individual bits.

The development of the interpreter evolved into a virtual processor for SCL. The interpreter has three registers: an accumulator, an instruction pointer, and a step-size register. It also has three flags: carry, negative, and zero. All instructions referencing the accumulator use the SCORPIO accumulator, while all instructions referencing or setting flags use the SCORPIO flags. This is another portable aspect of SCORPIO. The interpreter's registers are independent of the architecture used; therefore, the interpreter can be implemented on any system. This also reserves the actual registers of the microprocessor for use exclusively by the interpreter, which makes writing the interpreter much easier. Error checking is also performed by the virtual processor to ensure that the SCORPIO hardware is not damaged accidentally by an SCL program. A prime example where error checking is essential is the movement commands. If a leg is activated for too long, or not given enough time to return to its original position, the nitinol wire actuators can be severely damaged. In this case, only the interpreter controls how long the leg is "turned on" and also waits for the leg to return to its original state before executing the next instruction. Therefore, the hardware is protected from any accidental damage that could be directly caused by an SCL instruction.

A monitor program is necessary to control the main processes of the SCORPIO robot. It currently has four functions: load, go, enter interactive mode, and dump memory. It is the routine initially executed upon power-up of the device, and it initializes all of the registers, timers, and interrupts needed for successful operation. The monitor then waits to receive a command via the modem. The load command uses the XMODEM-CRC protocol to transfer programs from a PC to SCORPIO.[6] This protocol was chosen primarily for convenience, as data transfer packages using XMODEM-CRC are widely available for the PC, which is the development platform for all SCORPIO programs. In addition, XMODEM-CRC routines have been successfully implemented for microcontrollers in the 8051 family.[7] The go command executes an SCL program given its starting address. This command will place the starting address in the interpreter's instruction pointer. The SCORPIO interpreter is then invoked, which will execute the program and, when finished, return control to the monitor. The enter interactive-mode command is used to tell the robot to enter a state in which it can communicate with a PC via the modem. In this state, simple commands can be typed by a user at a PC; these commands will be sent to SCORPIO and executed immediately. This will eventually be used to send the go command; however, the 8051 emulator is currently used for process control. The final command is the memory dump, used only during interactive mode operations. This will let someone using the SCORPIO shell examine the contents of memory. The memory dump command will be useful in debugging SCORPIO when 8051 emulators are not available.

CONCLUSION

Many hours of diligent work have been invested in designing the software for SCORPIO, but there is still much to be finished. One planned improvement for the future is to write a SCORPIO shell that will enable a PC user to interact with SCORPIO. A tool of this kind will be extremely useful in testing and debugging SCORPIO control programs. Another possibility is routines to enable SCORPIO-to-SCORPIO communications via the IR modem. The flexibility of SCL allows additional hardware drivers to be added, such as robotic-arm controllers or ultrasonic vision devices, with relative ease. Further instructions, such as logic and arithmetic functions, and possibly instructions for manipulating one or more general-purpose registers, might also be added to broaden the functionality of SCL. Given the current design and implementation, future modifications can be achieved without modifying major portions of code. This enables the SCORPIO software to be a strong foundation for future additions or modifications, while at the same time providing a great deal of functionality.

REFERENCES

1. See Chapter 2.

2. Estell, J.K., and T.A. Owen. 1995. Experiencing the codesign process: Microcomputer Systems II laboratory. *SIGCSE Bulletin*, Vol. 27, No. 1: 34–38.

3. See Chapter 10.

4. See Chapter 2.

5. MacKenzie, I. 1992. *The 8051 microcontroller.* New York, N.Y.: Merrill.

6. Walrand, J. 1991. *Communications networks.* Homewood, Ill.: Aksen Associates, Inc.

7. Cameron, L.L. 1993. XMODEM-CRC on the 8031. *Computer Craft*, Vol. 3, No. 6: 61–65.

Chapter 12

Łukasiewicz' Insect: The Role of Continuous-Valued Logic in a Mobile Robot's Sensors, Control, and Locomotion

Jonathan W. Mills

The ability to physically realize a colony of insect-like robots presents numerous problems to robotics researchers. A hexapod robot controlled by a computational sensor is proposed as a solution to some of these problems. Stiquito is a small nitinol-propelled robot. It is controlled by a computational sensor implemented with Łukasiewicz logic arrays (ŁLAs). The computational sensor includes an electronic retina, an implicit controller, and a gait generator. Measured and simulated results illustrate the unifying effect of Łukasiewicz logic on the design of the robotic system.

Key words: analog VLSI
                Łukasiewicz logic array
                robotics
                subsumption architecture
                Stiquito

A reprint from an IEEE publication: *Proceedings of the International Symposium of Multiple Valued Logic*, Sacramento, CA, 1993.

INTRODUCTION

The Complex Insect

Research in robotics has advanced recently by lowering its expectations. Problems in path-planning, modeling, and image understanding were the focus of robotics research from 1960 to 1986. These problems are difficult in any domain, and have yet to be solved satisfactorily.

In 1986 Brooks proposed studying simpler robotic systems controlled by behavior-oriented rather than function-oriented networks. This approach is known as a *subsumption architecture*.[1] Attila and Genghis, two robots controlled by a subsumption architecture, resemble insects yet are capable of learning complex behaviors such as walking.[2] The concept of subsumption architecture has been extended to model the complete behavior of a cockroach-like robot,[3] and to the study of insect-like robot colonies.[4] But implementing a single insect-like robot is still difficult, and robot colonies are even more so. To understand why, let's consider the biological competition: the ant.

Ants are sophisticated creatures. They have eyes that implement lenses and filters using binary optics; they communicate both by touch and chemically using pheromones; they are strong for their size (capable by analogy of climbing and descending Mount Everest five times a day while carrying their own weight);[5] and they have multiple degrees of freedom in each leg, antenna, and the mandibles. By contrast, insect-like robots are clumsy and ponderous. Their optics range from simple photocells to television cameras; they cannot communicate using touch or chemicals; they are fragile and weak for their size; and they have only a few degrees of freedom in each leg or arm.

A single ant colony may contain up to 10,000 adults ranging in size from 3 mm to 25 mm in length. A colony has many castes specialized for specific tasks, yet capable of acting in concert to perform complex actions (bridging, wars, agriculture). Colonies of up to 20 robots have been implemented with "adults" approximately 500 mm in length that cost $2000 each. Of the robot colonies physically realized to date, none have castes designed for specific tasks, and concerted action is simple (such as flocking behavior). The weight-to-power ratio of small self-contained robots limits the functionality of their control and locomotion systems, which even then are difficult to construct.

The use of continuous-valued logic to implement a simple subsumption architecture for the small robot, *Stiquito*, is proposed as a step toward the solution of these problems.

Stiquito: Łukasiewicz' Insect

Stiquito is an insect-like hexapod robot propelled by nitinol actuator wires.[6] It is small (60 mm long x 70 mm wide x 25 mm high), lightweight (10 gm), and inexpensive ($10). Stiquito is capable of carrying up to 200 grams while walking at a speed of 3 to 10 centimeters per minute over slightly textured surfaces. Its payload typically consists of sensors, control and drive electronics, and a 9-volt cell. The robot walks when heat-activated nitinol actuator wires attached to the legs contract. The heat is generated by passing an electric current through the nitinol wire. The legs can be actuated individually or in groups to yield tripod, pacing, and other gaits.

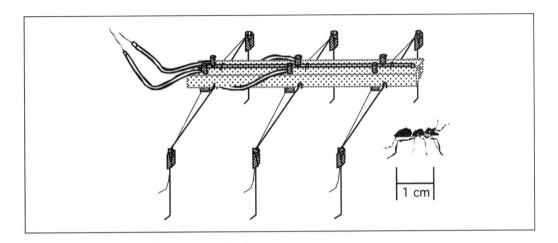

Figure 12.1. Stiquito and an ant (both x1).

Robotics and Continuous-Valued Logic

It is a paradox of robotics research that inherently imprecise and inaccurate biological systems are often implemented with digital systems far more precise and accurate than the original.

Digital implementations of hierarchical systems are flexible and easily programmable. However they weigh too much and consume too much power to be used to control an autonomous Stiquito. These systems are also unnecessarily complex. The programmability of digital microprocessors is useful during prototyping, but unnecessary in the final implementation; programmability can be emulated during design by simulation of the system. Conversion of analog sensor inputs to a digital representation and back to analog control outputs adds additional complexity that is not needed by an imprecise insect-like robot.

Consequently, a fully analog system was chosen to obtain lightweight, low-power, and simple sensors, controllers, and actuators, avoiding the paradox. Łukasiewicz logic arrays are used throughout the system. The continuous-valued Łukasiewicz logic provides a unified framework for the design of Stiquito's sensors, controllers, and locomotion.

Łukasiewicz Logic Arrays

Łukasiewicz logic is a multiple-valued logic with a denumerably infinite number of truth values.[7] Real circuits are described by subsets of the logic that have a finite number of continuously-varying truth values. No circuitry is required to quantize the logical values in the circuit. The limit to the number of values that can be encoded on a wire occurs when values cannot be distinguished because of noise.[8]

Łukasiewicz implication is defined by the valuation function $v(\alpha \supset \beta) = min(1, 1 - \alpha + \beta)$. *Negated implication* has a value function defined as $v(\alpha \not\supset \beta) = max(0, \alpha - \beta)$. The term *implication* is used to refer to both functions.

Łukasiewicz logic arrays (ŁLAs) are arrays of continuous-valued analog circuits for Łukasiewicz implication ($\supset$) and negated implication ($\not\supset$) circuits (Figure 12.2).

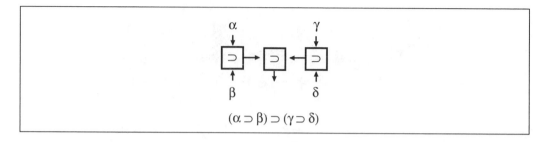

Figure 12.2. Fundamental ŁLA H-tree.

Negated implication yields the simplest circuit. It is composed of a current source, a current sink, a three-wire summing junction, and a diode-connected MOSFET. The output is accurate to within 1% of the full scale and precise to within six to eight bits.[9,10]

The dual logical and algebraic semantics of ŁLAs allows them to perform symbolic and numeric computations, a property essential to the application described in this paper.

SENSORS: THE ŁLA RETINA

Retina Design

A prototype ŁLA retina uses negated implication in a continuous-valued directed-edge sensor. This is a primitive vision function.[11] Four photocells are grouped in a cluster. Because negated implication is equivalent to a positive-only derivative, edges are detected in the NS, EW, SN, and WE directions across the cluster (Figure 12.3).

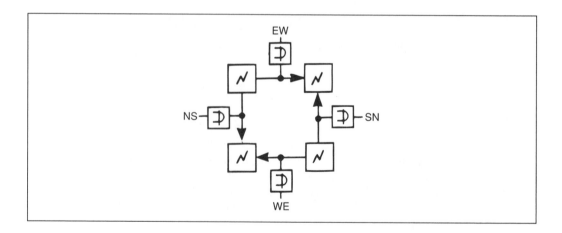

Figure 12.3. ŁLA retina four-pixel cluster.

While the ŁLA retina could be implemented with npn and pnp phototransistors and diodes,[12] our prototype uses npn phototransistors, MOSFET current mirrors, and diode-connected MOSFETs. It is still smaller and faster than Mead's retina[13] because the MOSFETs require less area than bipolar components (Figure 12.4).

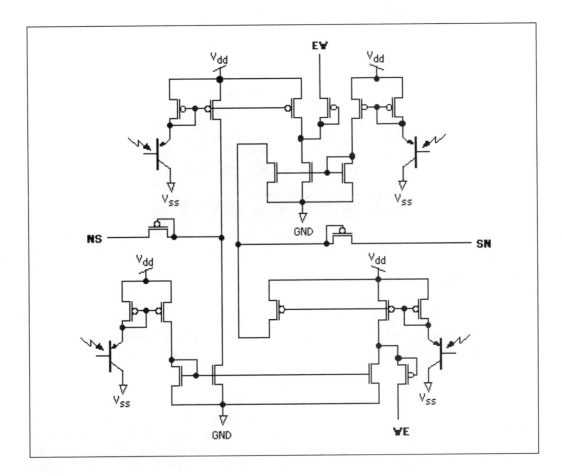

Figure 12.4. Four-pixel cluster schematic.

Prototype Retina EYE-1

An ŁLA retina prototype EYE-1 consisting of npn photocells and two edge detector strips was fabricated and tested to evaluate the design. Each photocell was laid out in a row consisting of other cells, each with the same size and shape. Two square photocells (bipolar transistors) were used, one 0.025 mm high x 0.025 mm wide, and the other 0.006 mm high x 0.006 mm wide (Figure 12.5).

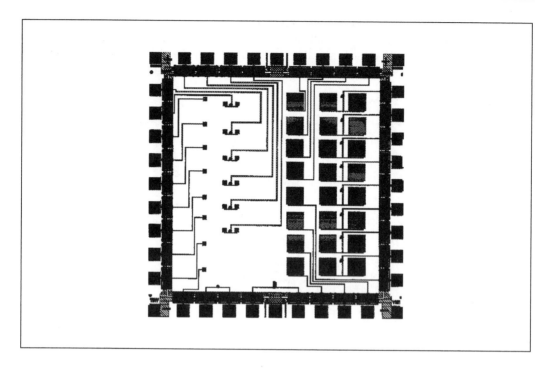

Figure 12.5. ŁLA retina prototype EYE-1.

The edge-detecting capability of the prototype was characterized by projecting an edge onto one of the photocells, then measuring the edge voltages computed by a circuit whose inputs came from all photocells in the row. The selected photocell's response was thus measured relative to the output of the other cells in its row.

An 8mm movie projector lens focused light onto the exposed window of the chip. To project a sharp edge a strip of wire was attached to the front of the light source so that its shadow could be swept across the chip surface. A simple amplification circuit was constructed with a voltage output. A photocell was darkened by the edge, then the edge was adjusted to obtain the maximum output value possible from that cell. The output of all photocells was recorded, then the edge moved to the next cell.

The outputs were most distinct when the smallest cells were tested (Figure 12.6). This is probably because the edge was slightly wider than the width of the photocell. Since the large cells are much closer together there was probably some overlap which reduced the edge detection. The smaller cells have about 10 times more space between them and were less prone to this overlapping phenomenon.

The retina is also fault-tolerant. It will continue to operate if a MOSFET fails in one of the sensor clusters, or if several clusters fail entirely. The output of prototype retina EYE-1 indicated that variations in the photocells due to fabrication affected their sensitivity, but did not prevent the devices from functioning.

Measurement of the phototype retina EYE-1 indicates that its minimum response time is limited by the response of the photocells rather than the negated implications. Photocells respond to edges in 20 ms, allowing all edges in an image to be identified in 20 ms. A 100-cluster retina could detect up to 5000 edge-crossings per second in a dynamically changing image. This is four orders of magnitude fewer than MOSFET negated implications can handle,[12] suggesting a direction for improvement.

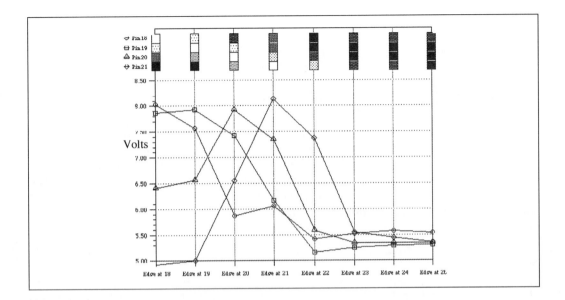

Figure 12.6. Output voltages from retina prototype.

The retina has some disadvantages. It is difficult to extract the results of the computation because the number of output pins is limited. If the data desired consists of a complete pixel map of edges, then each bit must be multiplexed to the output pins.

Another solution, used in the design of the controller, is to compact the results by merging computation with the sensor array. This type of device is known as a *computational sensor*. In this case, the edge data is compacted by computing the leg control signals, and outputting them instead.

Retina Simulation

The ŁLA retina was simulated using a spreadsheet program. Sensor clusters are arrayed to form a 10 x 10 retina (Figure 12.7b). Each cluster was implemented with eight spreadsheet cells, four for each photocell in the cluster and four to compute the edges detected by negated implication. A graph of the input photocells reconstructed the image (Figure 12.7a), with a second graph used to visualize the edges (Figure 12.7c).

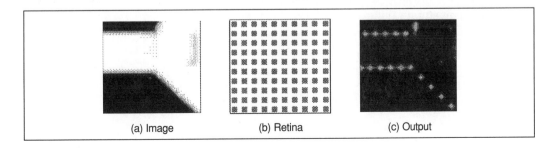

(a) Image	(b) Retina	(c) Output

Figure 12.7. Simulation of ŁLA retina.

The difference in pitch between horizontal/vertical clusters and diagonal clusters results in a stronger response to horizontal and vertical edges (Figure 12.7c). The retina also generates noise by detecting weak edges (such as the patch on the wall in Figure 12.7a).

Control: a Simple ⌐LA Subsumption Architecture

A simple subsumption architecture for navigating a maze was implemented by breaking the necessary behavior into simple tasks. The tasks included walking in a straight line, turning left, turning right, and avoiding the maze wall. Even these few tasks are sufficient to generate simple emergent behavior, such as escaping from corners and backtracking. The controller for this subsumption architecture is implicit. It is implemented by partitioning the sensor clusters into regions that correspond to the tasks in the subsumption architecture (Figure 12.8).

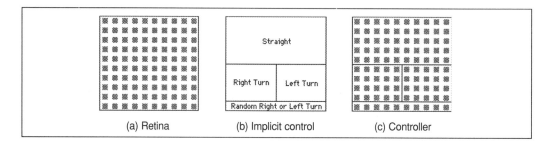

(a) Retina   (b) Implicit control   (c) Controller

Figure 12.8. Subsumption architecture controller.

Summing the edge signals in a given region results in an output signal for a specific task. The outputs are used to modify the robot's gait. Different tasks are activated as edges in the visual field move from region to region on the retina. A random left or right turn, a right turn, a left turn, and walking in a straight line compete for control of the robot based on the strength of each region's output. The random turn prevents the robot from getting stuck in corners and at walls. The robot's motion produces feedback that affects its behavior.

A simulation of Stiquito navigating a maze using a similar controller was developed for the Silicon Graphics Reality Engine.[14] Images from the simulation were digitized and transferred to the retina simulation. The sequence shows a robot moving through a maze (Figure 12.9a 1–5). As the image of the wall changes (Figure 12.9b 1–5) the edges move through regions in the controller (Figure 12.9c 1–5) to generate the tasks of the subsumption architecture (Table 12.1).

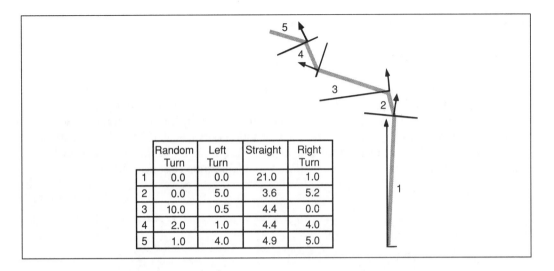

Table 12.1. Controller signals and path vectors.

	Random Turn	Left Turn	Straight	Right Turn
1	0.0	0.0	21.0	1.0
2	0.0	5.0	3.6	5.2
3	10.0	0.5	4.4	0.0
4	2.0	1.0	4.4	4.0
5	1.0	4.0	4.9	5.0

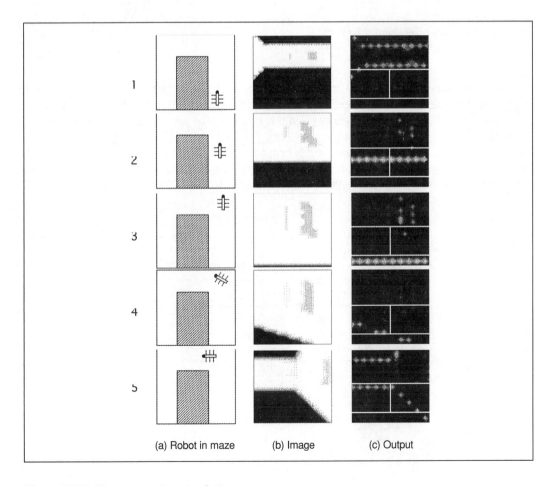

Figure 12.9. Maze navigation simulation.

LOCOMOTION: THE ŁLA GAIT GENERATOR

Individual Leg Chaotic Controller

Łukasiewicz logic can be used to specify chaotic systems.[15] By assigning truth values to components of a dynamical system Łukasiewicz logic can describe and control the behavior of the system. The ŁLA gait generator is constructed out of six identical chaotic controllers, one for each leg. The operation of a chaotic controller is defined by four functions controlling the trajectory of the foot: Backward ( ), Forward ( ), Raise ( ), and Lower ( ). The x- and y-motions are expressed as difference equations (Figure 12.10).

$$x_{t+1} = x_t + \Delta x_{t+1} \;,\; \left\{ \Delta x_{t+1} \;\middle|\; \begin{array}{ll} x_t > y_t & \text{Backward}(x_t) \\ x_t \le y_t & \text{Forward}(x_t) \end{array} \right\}$$

$$y_{t+1} = y_t + \Delta y_{t+1} \;,\; \left\{ \Delta y_{t+1} \;\middle|\; \begin{array}{ll} x_t > y_t & \text{Raise}(x_t) \\ x_t \le y_t & \text{Lower}(x_t) \end{array} \right\}$$

Figure 12.10. Self-referential sentences.

The difference equations are mapped to Łukasiewicz logic arrays, and correspond to self-referential sentences of Łukasiewicz logic. Although the four functions may be interpreted as fuzzy functions, using fuzzy linguistic modifiers to define them did not work. Rule sets that combined functions such as *leg very far back* and *leg somewhat raised* ended up producing "stiff" leg trajectories with abrupt changes. When a smooth hand-drawn path was quantized, and the four resulting rules applied, a much improved trajectory resulted. The rules are shown graphically, scaled to the range –0.3 to 0.3 (Figure 12.11).

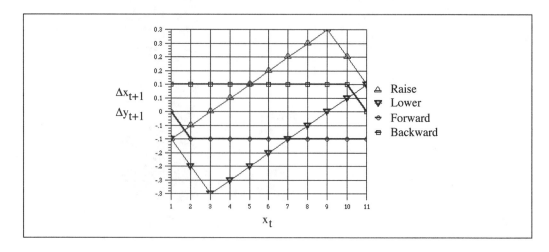

Figure 12.11. Leg motion rules.

The rules do more than define a single trajectory. Because the rules define relative changes in the x and y axes, they apply throughout the space of truth values to define a state space containing a family of limit cycles. The limit cycles define multiple trajectories for the robot's foot in the state space (Figure 12.12).

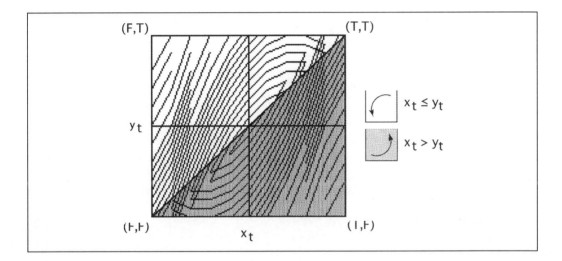

Figure 12.12. Limit cycles in the state space.

A trajectory is stable if not perturbed (Figure 12.13a). Small perturbations, due for example to shot noise in the ŁLA circuit, will cause the trajectory to "jitter" (Figure 12.13b). Perturbations above a threshold $\approx \frac{1}{15}T$, such as a variable control signal applied as an input, will lead to increasingly chaotic behavior as trajectories alternate (Figure 12.13c).

Because the rules are evaluated algebraically, their effect extends outside the space of truth values if *min* and *max* are omitted from the valuation functions of logical connectives. This was done in spreadsheet simulations of the chaotic controller and the gait generator to view unclipped trajectories.

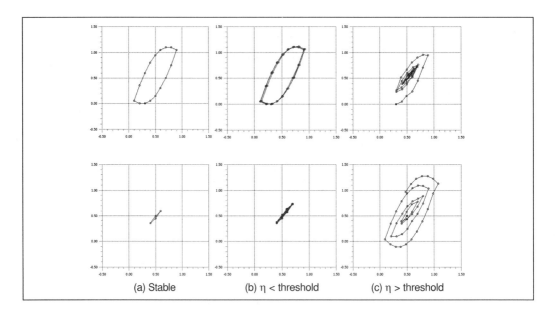

Figure 12.13. Limit cycles perturbed by noise η.

The Gait Generator

The gait generator for Stiquito is constructed from six chaotic controllers, one for each leg. The leg controllers are implemented with negated implication and packed into the unused space of the ŁLA retina. The computational sensor merges the ŁLA retina and the gait generator, and fits onto a MOSIS Tiny Chip (Figure 12.14).

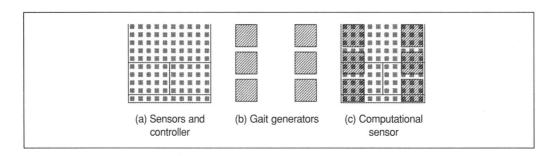

Figure 12.14. ŁLA subsumption architecture implemented as a computational sensor.

Because Stiquitos built to date have only one or two degrees of freedom per leg instead of four, two or three of the outputs of each chaotic controller can either be ignored or wire-summed to combine control in the x and y axes. Even so, the gait generator will produce a wide variety of gaits. Because the trajectories do not switch synchronously, out-of-phase oscillations in the unsynchronized trajectories produce many variations of the full tripod gait (Figure 12.15a).

Gaits that approximate single leg movement arise when a weak oscillation in five legs is coupled to a strong oscillation in a single leg. This is similar to the weak motion of the left legs shown in Figure 12.15b. Such a gait can be generated spontaneously by applying a strong perturbation to all legs until the trajectories are out-of-phase and degenerate. This is akin to many gait generators.[16]

Turning gaits result when several legs on one side have strong oscillations, while legs on the opposite side have weak oscillations (Figure 12.15c). A turning gait is obtained immediately by forcing the legs on one side into out-of-phase and degenerate trajectories.

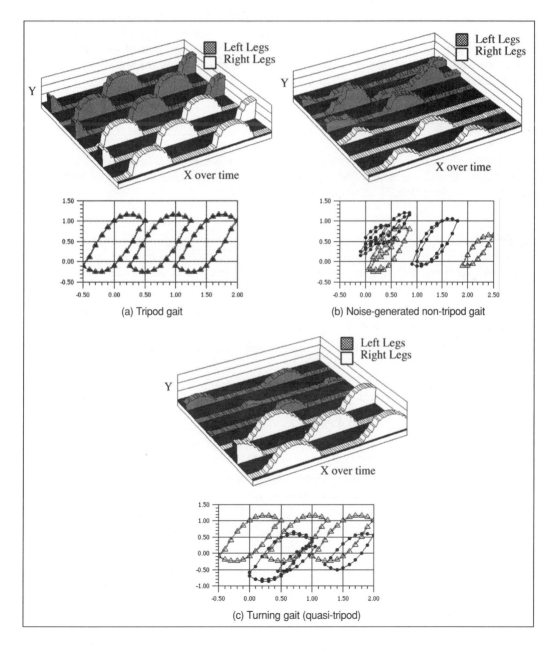

Figure 12.15. Gaits.

SUMMARY

Łukasiewicz logic unifies the design of sensors, control, and locomotion for a small, simple robot. The simplifications that result solve problems of weight, power, and complexity in the design of insect-like robots. The next step is to fabricate the computational sensor described here on a MOSIS Tiny Chip, and use it to control Stiquito.

ACKNOWLEDGMENTS

This research was funded in part by NSF Grant MIP-9010878. Andrew Heininger fabricated the ŁLA retina prototype EYE-1; he and Amitava Biswas tested it. Shyam Pullela fabricated a programmable FPGA gait generator. Paulo Maciel developed an SGI graphic Stiquito simulator to study vision and machine learning in simple mazes. Jason Almeter suggested improvements to the implicit controller.

ACCESS TO THIS RESEARCH

Instructions and parts lists to construct Stiquito; simulators for Stiquito, the ŁLA retina and gait controllers; and design files for FPGA controllers are available by anonymous ftp from cs.indiana.edu (129.79.254.191) in /pub/stiquito.

REFERENCES

1. Brooks, R. 1986. A robust layered control system for a mobile robot. *IEEE J. Robotics & Automation* **RA-2**(1).

2. Brooks, R. 1990. A robot that walks: Emergent behaviors from a carefully evolved network. *Neural Computation* **1** (2): pp. 253–262.

3. Beer, R. 1991. An Artificial Insect. *American Scientist* **79** (September-October): pp. 444–452.

4. MacLennan, B. 1990. *Evolution of Communication in a Population of Simple Machines*. University of Tennessee Computer Science Department TR CS-90-99.

5. Sanderson, I. 1965. *Living Treasure*. NY: Pyramid.

6. Mills, J.W. 1992. *Stiquito: A small, simple, inexpensive legged robot*. Indiana University Computer Science Department TR 363a. See also Chapter 2.

7. Łukasiewicz, J., and A. Tarski. 1930. Untersuchungen über den Aussagenkalkül. *Comptes rendus des séances de la Société des sciences et des lettres des Varsovie Classe III* (23): pp. 30–50.

8. Montante, R.A., and J.W. Mills. 1993. *Measuring Information Capacity in a VLSI Analog Logic Circuit*. Indiana University Computer Science Dept. TR 377.

9. Mills, J.W., M.G. Beavers, and C.A. Daffinger. 1990. Łukasiewicz Logic Arrays. *Proceedings of 20th International Symposium on Multiple-Valued Logic*.

10. Mills, J., and C. Daffinger. 1990. CMOS VLSI Łukasiewicz Logic Arrays. *Proceedings of Application Specific Array Processors.* Princeton, New Jersey.

11. Ballard, D.H., and C.M. Brown. 1982. *Computer Vision.* Englewood Cliffs, NJ: Prentice-Hall, Inc.

12. Mills, J.W. 1992. Area-Efficient Implication Circuits for Very Dense Łukasiewicz Logic Arrays. *Proceedings of 22nd International Symposium on Multiple-Valued Logic.* Sendai, Japan: IEEE Press.

13. Mead, C. 1989. *Analog VLSI and Neural Systems.* Reading, Massachusetts: Addison-Wesley.

14. Maciel, P. 1993. Stiquito simulator for SGI. See also Chapter 15.

15. Stewart, I. 1993. Mathematical Recreations. *Scientific American.* February 1993: pp. 110–112.

16. Donner, M.D. 1987. *Real-Time Control of Walking.* Boston, Massachusetts: Birkhäuser.

Chapter 13

Stiquito, a Platform for Artificial Intelligence

Matthew C. Scott

INTRODUCTION

This chapter presents examples and opportunities for studying and applying the paradigms of artificial intelligence (AI) using the simple and inexpensive platform provided by the Stiquito robot.

- The mechanics of articulated hexapod ambulation are inspected and a genetic algorithm is presented for determining the optimal gaits that Stiquito may attain while minimizing energy consumption and maintaining balance.
- An algorithm is presented for controlling the legs that allows for more than $\binom{240}{3}$ forward-propelling gaits. Its development on a computer-aided design system and implementation on a field-programmable gate array (FPGA) are illustrated.
- The algorithm is then presented as a neural network using chains of phase-locked loops, and implementation using CMOS (complementary metal-oxide semiconductor) VLSI (very-large-scale integration) technology is investigated.
- Some possible experiments are considered for producing interesting populational dynamics, such as emergent cooperation and self-organization.

In light of the educational aspect of Stiquito's motivation, the controller described in this chapter is pursued with a distinct bent toward its extension to possible term projects in AI, robotics, digital design, and VLSI courses. But the primary objective is the development of a VLSI neural controller. Only a cursory knowledge of the tenets of neural nets, genetic algorithms, LSI (large-scale integration) and VLSI design is necessary to appreciate the ideas and designs presented.

INSTANTIATING ARTHROPODAN AMBULATION

Everyone, no doubt, has spent some time marveling at the dexterity, strength, and speed of the common ant, not to mention the obviously complex behavior required to negotiate the rugged terrain to which this creature is accustomed. The complexities imposed by nonuniform surfaces, turning, backing up, and damage will not be addressed here, since providing for these characteristics makes the project intractable to a semester course and requires some interesting additions to Stiquito's architecture, as can be seen in Raibert.[1] But this has certainly not tempered the ambitions of Stiquito's proponents, as the ability to experiment with such complexities was one of the primary motivations for Stiquito's conception. Thus, to produce simple walking, the controller discussed in this chapter is required only to determine the sequence and phase periods of signals being sent to the individual leg actuators. The coordination required for also chewing gum will be relegated to a future project.

Observation has shown that insects employ several different gaits at various speeds.[2,3] There are many methods by which a controller may be designed to produce the desired gaits. Implementing the controller using a binary logic entails the development of a traditional algorithm state machine (ASM). Given a fuzzy logic, a similar fuzzy ASM is needed. But such a fuzzy ASM, being stochastic in nature, does not surrender itself easily to verification. Yet it is suspected that a fuzzy controller would be preferable as it could provide an acceptable, or at least smoother, behavior in the transitions between gaits. It is also believed that expansion of the algorithm to produce the aforementioned complex behaviors will require much less additional hardware with the fuzzy-logic implementation than with a binary logic. This belief is based on the inherent brittleness of traditional digital von Neumann machines and their typical resistance to extrapolation of state-meaning to the multiplicity of possible environmental conditions. Each element of a fuzzy system, on the other hand, can itself be seen as a substate-machine whose number of states is bounded only by the number of values that an element in the logic may assume. Thus, given n fuzzy-logic elements of order s possible states and an integrating or differential network with a k degree of connectivity, the system can assume $kn^s!$ states. Of course, a general-purpose computer can be programmed to simulate a fuzzy system, but then the advantage of parallelism is lost and information that may be embedded in the time-dependent relationships of asynchronous logic elements and real-time inputs is virtually impossible to recover.

Fuzzy logics can be seen as constrained neural networks (NNs) wherein each neuron has (minimally) two inputs and one output. Such an architecture lends itself admirably to a Łukasiewicz Logic Array (ŁLA) implementation. ŁLAs are

> *...massively parallel analog computers organized as binary trees of identical processing elements performing either implication ($\Longrightarrow$), negated implication ($\nRightarrow$) or both.*[4]

and

> *By showing that valuation functions for connectives in sentences in Ł are equivalent to piecewise-constructable first-degree polynomials that map the hyperspace $[0,1]^n$ into the interval $[0,1]$, the capability for building fuzzy pattern recognizers is provided.*[5]

The ŁLA, in simpler terms, derives an implication function by the differentiation of two electrical currents. An organization of H-tree cascaded current mirrors and Schottky diode towers accepts a sentence in balanced normal form of implication and computes a bounded difference function according to those inputs. Considering the limitations of VLSI design, this may be the most promising technology for implementing such a fuzzy controller. Recent developments allow the placing of 36,000 ₤LA implications in the 1,800 μ by 2,000μ area of one MOSIS tiny chip.[6] Such an unprecedented yield effectively places the motif of AI in the applications arena. But because of its current mode-analog computation and Kirchoff's laws, instantiating recurrency is prohibitively complex. In this respect the ₤LA seems best suited for feed-forward nonrecurrent NNs. There are, however, various strategies for circumventing this limitation, which are currently under investigation. Also found is the suggestion of using double-poly capacitors as storage elements for weights.[7] A similar approach is presented by Carver Mead[8] where an RC circuit acts as a differentiator and a transfer function.

The next section of this chapter describes an NN layout whose architecture follows the propositions of both Mills[6] and Mead.[8] An algorithm is presented that requires chains of phase-locked neural loops and, thereby, a high degree of recurrency. It is possible that such loops are integral to the operation of the mind. In certain respects, this belief is borne out by the successes of the subsumption architectures propounded by Brooks et al.[9] The idea has motivated the development of a VLSI NN implementation that might realize such a hypothesis.

To envision how loops might be biologically grounded, consider the typical human's affinity for music. It may be the case that the periodic structure of music maps well to periodic subconscious cycles in the brain. This periodic structure manifests itself in the human tendency to repeat a few bars of a tune over and over inside one's head. This natural noumenon lends credence to the theories of associative memory models[10] wherein the activation of the mental state associated with one event causes the succeeding activation of another state. The assertion of these state relations has the effect of reinforcing their associations (that is, Hebbian learning). If this is true, then the periodic nature of music may facilitate the streamlining of the brain's operating system, and thus receives positive reinforcement. Brainstorming, on the other hand, is often perceived as painful because it requires, as its name implies, an agitation of mental processes.

It should be noted that the hypothesized cycles of the mind are not likely due to hardwired loops, but rather repeating patterns of state of mind. And, unlike a Markov chain, which is bound to recur eventually, the mind, inherently neuromorphic, is highly unlikely ever to recur exactly to a previously held state. But it will, because of the basins of attraction formed by the connection weights, tend to oscillate within a bounded region of the brain's state-space. The reader might recognize this as the incarnation of that ubiquitous beast of chaos, the Strange Attractor. Note also that the mind is likely composed of many intertwined loops of various degrees of strength and period. New loops are constantly being spawned and others dwindle away.

The concept of loops is also supported by the incidence of gamma-wave oscillations induced by the six-layer recurrent networks in the columnar laminae of the visual cortex. It has been proposed[11] that the phase locking detected between disparate columns provides a foundation for solving the binding problem between visual percepts and conscious awareness. For an excellent source of readable papers on the subject of coupled oscillating neural nets and their analysis as pattern generators, the reader is referred to Taylor and Mannion.[12]

With this in mind, the digital ASM has been tailored to model (albeit, not in the distributed sense) the intended behavior of the NN implementation. That is, the ASM is actually 12 separate ASMs with the data dependencies between each being limited to a single flag. Each subASM constitutes a cycle that emulates the proposed loops of the NN implementation. The cycles are linked together as a chain, and form two arms of a spanning tree structure (see Figure 13.1). The root of this tree is the master ASM. Further explanation is deferred to the next section. But it is important to realize that the excessive logic required for this digital implementation is necessary to provide a common structure between it and

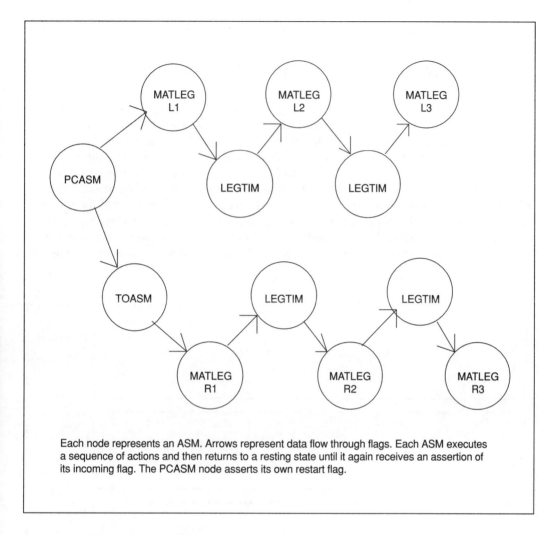

Each node represents an ASM. Arrows represent data flow through flags. Each ASM executes a sequence of actions and then returns to a resting state until it again receives an assertion of its incoming flag. The PCASM node asserts its own restart flag.

Figure 13.1. Structure of the ASM.

the VLSI NN version. This allows comparison of the sizes, speeds, and relative fault-tolerance capabilities of the two systems. The looping structure has also created the unintended consequence of making the digital controller very versatile in terms of the number of possible gaits derived from a few variables.

The minimal circuitry required to control an insect's locomotion need only include the inputs of leg position and desired speed. The controller described here does not require the inputs of leg position as these are determined internally, relying on the consistent periodic behavior of the legs. That is, it is known how long it takes for a leg to flex and relax due to the momentary heating of the nitinol wire. With this knowledge the algorithm should be made to allow ample time for a leg to complete its "kicking" phase before starting it in another cycle. The algorithm should be optimized with respect to the order and frequency of the current pulses being sent to the leg's nitinol actuators. This optimization will include the following:

- Minimizing the energy required to maintain a certain speed
- Maintaining balance
- Providing a symmetric and cyclic pattern of leg movements to minimize the logic required

The parameters of balance and the cyclic pattern seem to be directly related in that the symmetry between the behavior of the left- and right-hand sides in stable gaits tends to provide a smooth and constant propelling force. As the speed increases the patterns seem to increase their offset (see Figure 13.2). It is possible that this is because the period of the cycles at greater speeds is shorter and thus there is less time for the unsupported side to fall. Above a certain speed the falling time between leg impulses may be so small that no organized cyclic pattern is needed; the behavior becomes chaotic. But it may be the case that the insect's neural system is not capable of computing beyond a certain rate. This is reasonable in that a neuron has a limited firing rate and the neurotransmitters have a constant rate of propagation along the axons and dendrites. Above a certain speed it may not

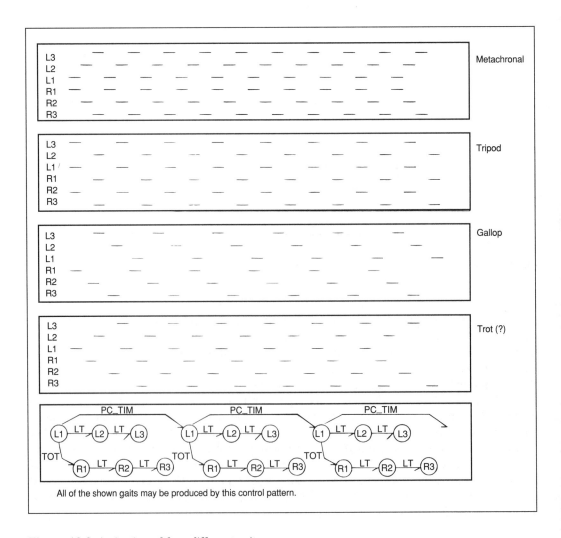

Figure 13.2. Activation of four different gaits.

be feasible for a leg-jerk neuron to wait for the computing circuitry to determine whether it should fire, because in that time it should have fired many times already. In any case, a silicon controller will not likely need to employ chaos, as its speed of processing is several orders greater than the reaction time of the nitinol. So, in developing the control algorithm a higher fitness value should be given to sequences that are cyclic and regular and tend to provide symmetrical support about the center of gravity. This symmetry is apparent in nature, in the form of the tripod and metachronal wave gaits. Another constraint that may be considered is whether the legs on one side will interfere with each other (that is, a middle leg back and a back leg forward). This implementation does not consider this to be the case, but it may be relevant in future designs.

To minimize the energy function it can be expected that an optimized algorithm will evenly disperse the leg thrusts on each side of the robot in a pattern that minimizes friction from dragged legs and minimizes leg extensions, wherein the work necessary is done by one leg and the others are a useless redundancy.

To actually define the representation of the controller on the genetic algorithm (GA) string it will be necessary first to describe the development of the controller and its required input values.

A SIMPLE YET VERSATILE DIGITAL CONTROLLER

This section describes a possible algorithm for inducing walking in the Stiquito robot. The digital logic is developed with ViewLogic for implementation on an Actel Field Programmable Gate Array System. First a breakdown of the required time parameters is presented. Then a digression is made to motivate the genetic algorithm optimization. A quick sketch of the logic required is then presented, although excessive detail is not provided as it is believed that the general idea is well developed in this chapter and documentation sufficient to reproduce the design can be found in the figures and digital layout schematics. The ViewLogic design data files and more can be obtained from http://ece.arizona.edu:8001/~scottm.

It can be inferred from Figure 13.2 that all of the stable gaits of insects employ variations of metachronal waves of the legs on each side of the body. That is, it seems that the legs always kick in the sequences of L1-L2-L3 and R1-R2-R3. It follows that all of the gaits can be produced with a tuple of four time periods, which are as follows:

1. The time between the beginning of one sequence and the beginning of the next sequence on the same side of the body. This is called the PC (primary cycle) time.

2. The time between the beginning of a wave on one side of the body and the beginning of the corresponding wave on the other side of the body, called the TO (time offset) time.

3. The time between one leg kicking and its following leg kicking. This is called LT (leg-interval time).

4. Possibly, the duration of one leg extension cycle. This will probably be constant as the current driven through the nitinol wires is constant, and varying the duration does not vary the rate or degree of contraction. It is referred to as TK (time of kicking). In implementations using some variable force for propulsion it can be used to control the speed of the robot.

These time parameters, with the inclusion of a categorizing speed tag, may represent the GA genotype of this simplified insect. The phenotype is obtained by applying these terms to an objective function that computes for the characteristics to be optimized as discussed previously. It is believed that the GA will perform better in finding an optimal configuration than other heuristics because the problem space is presumed to be fairly regular with continuous gradients and possibly many local minima.[13] This can be seen intuitively by noting that the time constants are effectively continuously valued and lack physical dependencies.

Figure 13.2 also depicts the means by which the first three time parameters can be used to drive the robot. The parameter for current-pulse-frequency (TK) is not necessary but is still included in the design so that it may be modulated for different actuators of various resistivities and power requirements. The leg sequencing circuitry is based on six counters and five registers. One counter will determine the frequency to set the PC flag, another will control the TO flag, and all six will be used to sequence the legs on either side of the body by the setting of LT flags. The time-tuples for each of the four described gaits are downloaded to the controller by latching them into a register and then, via a multiplexor, into corresponding time registers. These registers are then referenced by the various segment's counters and comparators to determine when a phase or cycle should begin.

The discerning reader may have noted that a more hardware-efficient method would require only one counter, eight registers, a 3-to-8 multiplexor, and seven comparators. The controlling computer would simply download to the robot the time required for one full cycle of leg kicks and six time constants, each representing the time at which one of the legs should kick. The design presented in Heininger and Hendry[14] is particularly well suited for such an implementation. The extra parts required can easily be added to the design in order to produce the entire controller on one chip. Again, hardware efficiency is not the primary goal of this implementation.

It is believed that with this design the legs will not degrade into chaotic kicking due to undefined states because they will naturally attain a state of rest until signaled by a valid control signal. That is, each leg will follow its ASM, which will eventually leave it in a resting state until it again receives a kick signal (see the ASM diagrams included with each of the ViewLogic schematics). Because of the spanning tree chaining of the ASMs, the whole system is bound to come to rest. Unexpected states might only arise if the root ASM becomes unstable or if signals were able to propagate up the chains.

IMPLEMENTING THE CONTROLLER AS A CHAIN OF PHASE-LOCKED LOOPS

Figure 13.3 shows a sketch of a neural topology that follows the architecture and algorithm of the digital ASM. The importance of this representation is its possibility of being produced with phase-locked loops in a recurrent NN. Note that this is a symbolic representation. The initial VLSI NN can be hardwired with discreet loops to simulate the desired behavior, but the final implementation will entail a subsymbolic distributed architecture. The discreet, hardcoded looping constructs have been designed and tested with a homemade scheme NN implementation. It appears that they are sufficient for the production of cyclic patterns in an NN. Of course, many more neurons will be required than are shown in Figure 13.3 in order to achieve the proper timing. The interesting part will be to train a fully distributed NN to produce the desired cyclic behavior complete with subcycles and resting states. Inspired by various works,[15,16,17,18,19,20,21,22] a simple VLSI implementation has been attempted to do exactly that.

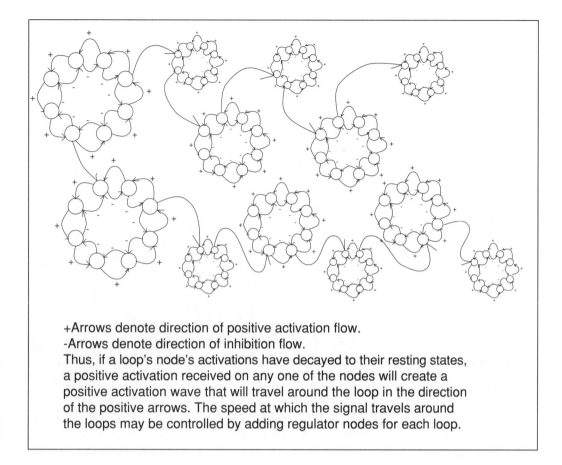

+Arrows denote direction of positive activation flow.
-Arrows denote direction of inhibition flow.
Thus, if a loop's node's activations have decayed to their resting states,
a positive activation received on any one of the nodes will create a
positive activation wave that will travel around the loop in the direction
of the positive arrows. The speed at which the signal travels around
the loops may be controlled by adding regulator nodes for each loop.

Figure 13.3. Neural topology of the digital ASM.

Consider that each of the approximate 10^{11} neurons in the typical human brain has, on average, connections to 10^5 other neurons. Even if it were reasonable to fit 100,000 neurons (say, 100 microns square each, in size) on a chip, it would still be practically impossible to implement a degree of connectivity approaching that of the human brain. The current state of the presented VLSI NN layout (see Figures 13.4, 13.5, and 13.6) attempts to minimize the area required for these connections by having one line, or axon, per neuron that proximally visits all the other neurons that provide inputs into it, the integrating neuron. That is, the dendritic tree of each neuron is collapsed into one line. Another strategy that will likely be employed is structuring the neurons into groups (ganglia) or cellular automata. That is, a group will consist of a number of nodes connected in either a random or hypercube topology. The groups themselves will be similarly connected.

The other major problem in the design of this analog NN has been in keeping the integration and output functions isolated (required for recurrency). This is currently accomplished with the use of a two-phase latch (the equivalent of a synapse). If the input to this latch reaches a certain threshold (3 volts), the latch will fire (output a value equal to Vdd), otherwise its output will be GND. In order to have a variable strength (analog) output from the neuron, the output of the latch switches a gate whose source provides a voltage whose value represents the weight of that neuron's departing axon. The training logic must vary

Electrical schematic of the VLSI neuron shown in Figure 13.5.

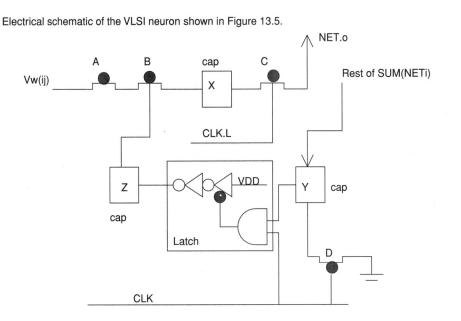

Vw(ij) represents the voltage weight of the pulse generated by this neuron when it fires. The boxes represent capacitors, or regions of double-poly layers which may be made to hold charges. The two inverters and the AND gate represent the latch. If the activation charge in the Y cap is large enough, latch will toggle, causing the Z cap to be charged by VDD. When CLK goes high the voltage of Vw(ij) will induce charge accumulation in X if Z acquired a charge before. When the CLK signal goes low, any charge accumulated on the X cap will be drained into NETo. Gate D is necessary to clear the input Net between cycles.

Topology of the network of 4 neurons shown in Figure 13.6.

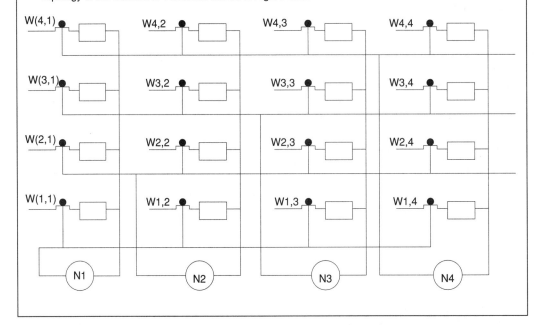

Figure 13.4.

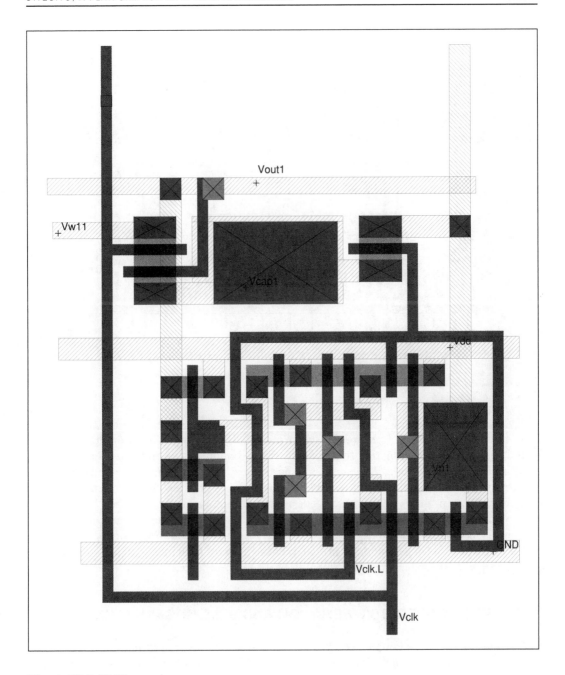

Figure 13.5. VLSI neuron.

this weight source. Regretfully, this logic has proven to be most untenable. But possibilities abound: A promising strategy requires that the NN be composed of two templates or sheets; that is, one fully functional NN on one sheet and an overlaid sheet whose functionality maps to the primary sheet. The word *sheet* is used instead of *layer* to prevent confusion with the traditional meaning of layer in the NN vernacular. The primary sheet would be the actual operating system and the other would record the activity and possibly the

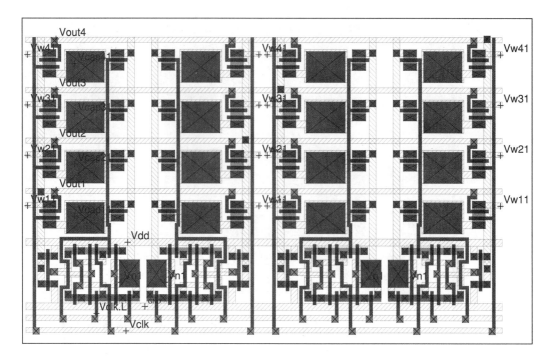

Figure 13.6. Network of four neurons.

success of the primary. Intermittently the primary sheet's connection weights would be modified according to a function that considered the conditions of both sheets as a fitness measure of recently produced behaviors of the primary net. Note that this allows incorporating the ideas of backpropagation and Hebbian learning. It is also important to note that such an architecture may be seen as a good model of biological systems. That is, real neurons tend to contain many different and somewhat independent neurotransmitters. In a certain respect, these transmitters can be seen as representing different copies (or sheets) of the same net. But the important property of such an architecture is that it enables the net to evolve at the same time it is engaging real-time problems.

EMERGENT COOPERATIVE BEHAVIOR AND SELF-ORGANIZATION

The development of an interface such as is found in Heininger and Hendry[23] creates the possibility of addressing and controlling up to 128 Stiquito robots. Such an army could demonstrate some interesting attributes of group dynamics. To conceive of a means of producing cooperative behavior, a mental model must be made in which each Stiquito is simply a coordinate point in a plane. There must be a causal relationship between these points such that they can affect each other and possibly develop complex structures or behaviors based solely on their relative positions, and possibly the discovered positions of obstacles or goals. But without sensors each robot is essentially enveloped in its own universe. Each is required to have some means of sensing (ultrasonic, infrared) whereby it can at least discover the presence of objects directly in front of it. Each of the robots may also be armed with a receiver/transmitter system for detecting the proximity of other robots. It would be advantageous to have four directional receivers (phototransistors) such that the robots could determine the direction of any nearby companions. Given a central controller and a Stiquito configuration as just described, some interesting and possibly useful cooperative behaviors can be realized.

- *Foraging.* Not exactly an emergent behavior, but potentially useful and easily done. The robots simply maintain a formation and walk through a field. Collisions with objects are minimized by incorporating fish-school tactics. The robot schools could be useful in any sort of work requiring the coverage of a large tract of land.

- *Terrain mapping.* Each robot builds a local map (a navigational template[24]). The combination of the local maps forms a global map. But a robot's local map cannot contribute to the global map until the robot knows its general position on the global map. When this is determined, it can match or patch its local map to the global map and reference the global map to determine a global strategy for obtaining an objective.

- *Robotic field games.* The Stiquitos, in the current design, are not likely to cause much discomfort to anything bigger than a microbe. (Actually, the name *Stiquito* is becoming synonymous with *touché!* It seems no one who has determined to build a Stiquito robot has avoided getting stung—by the insertion of piano wire into a finger. It is recommended that builders proceed with ample light and a clear, decaffeinated mind.) But nonviolent positional battles can be found in many board games wherein the pieces represent differing strengths and abilities (such as chess, tank battle, or Risk). To introduce a greater degree of complexity into the system, each Stiquito can be assigned various parameters such as weapons complement and strength. Each of the competing armies of Stiquitos is then given the task of configuring their formation and movement strategies to overcome the opponent in some respect. This has innumerable possibilities. The objective is to create rules and parameters such that an emergent cooperative behavior can be detected in the positions, movements, and status LEDs of the robots.

SUMMARY

Interest in the opportunities and potential provided by the recent discoveries and developments in AI has grown phenomenally during the last decade. The advent of high-powered PCs and workstations has enabled millions of people, from high-schoolers to Nobel laureates, to engage in serious, leading-edge research and development. But there has been a great propensity for the heralded theories and simulations to go no further than that, just theories and simulations. The Stiquito robot provides an opportunity for many of these ideas to be implemented and tested under the most strenuous and demanding of environments: the hardware lab and the common garage. But, as this article purports, the robot is also an excellent academic tool for students to apply to various AI projects. This article has presented some of the paradigms of AI that have come under consideration in developing a VLSI neural controller to produce dynamic and flexible hexapod locomotion. It is expected that this objective will soon be borne out by the Stiquito robot.

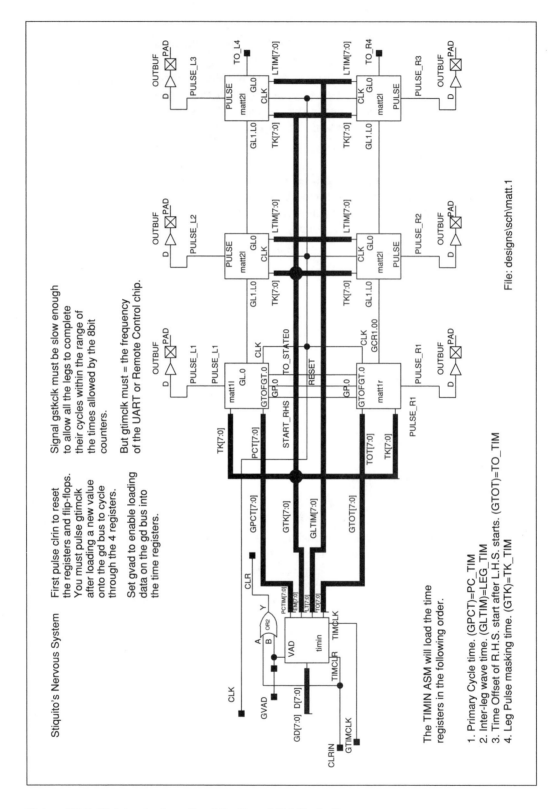

Figure 13.7. High-level schematic of the Neural Net Controller.

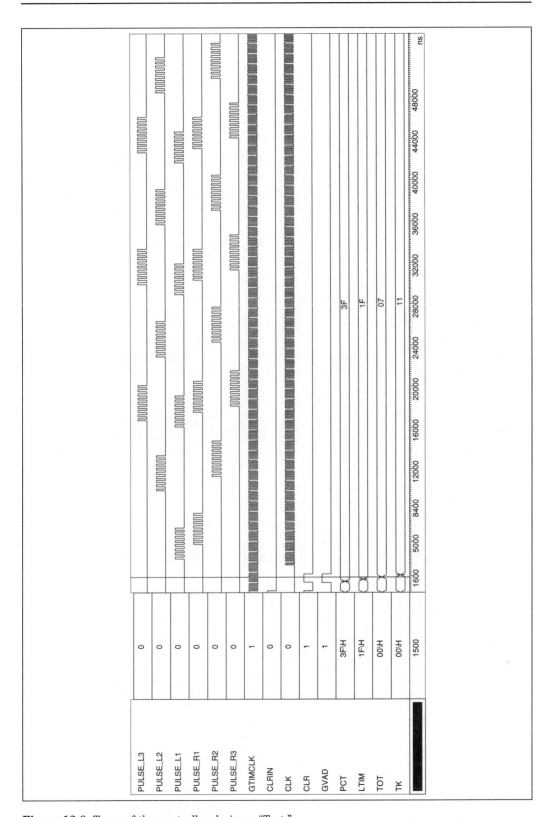

Figure 13.8. Trace of the controller during a "Trot."

REFERENCES

1. Raibert, Mark H. 1986. *Legged robots that balance.* Cambridge, Mass.: MIT Press.

2. Beer, Randall D., Hillel J. Chiel, Roger D. Quinn, Ken Espenschied, and Patrick Larsson. 1992. A distributed neural network architecture for hexapod robot locomotion. *Neural Computation*, Vol. 4, No 3.

3. Wilson, D.M. 1966. Insect walking. *Annual Review of Entomology*, No. 11: 103–122.

4. Mills, Jonathan W., M. Gordon Beavers, and Charles A. Daffinger. Łukasiewicz logic arrays. Technical Report No. 296, Indiana Univ., Bloomington, Ind.

5. Mills, Jonathan W., and Charles A. Daffinger. An analog VLSI array processor for classical and connectionist AI. *Proc. of Application Specific Array Processors.* Princeton, N.J.: IEEE Press.

6. Mills, Jonathan W. 1992. Area-efficient implication circuits for very dense Łukasiewicz logic arrays. *Proc. of 22nd Int'l Symp. on Multiple-Valued Logic.* Sendai, Japan.

7. Ibid.

8. Mead, Carver. 1989. *Analog VLSI and neural systems.* Addison Wesley.

9. Brooks, Rodney A. The role of learning in autonomous robots. *Ninth International Conference.*

10. Hinton, Geoffrey E., and James A. Anderson. 1981. *Parallel models of associative memory.* Lawrence Erlbaum Associates, Inc.

11. Crick, F., and C. Koch. Towards a neurobiological theory of consciousness. *Seminars Neurosc.*, Vol. 2: 263–275.

12. Taylor, J.G., and C.L.T. Mannion (eds.) 1992. *Coupled oscillating neurons.*

13. Rawlins, Gregory J., ed. 1991. *Foundations of genetic algorithms.* Morgan Kaufmann Publishers.

14. Heininger, Andrew, and Slamet Hendry. 1992. Parallel 2 serial controller. Indiana University, Technical Report No. 363, Part 2.

15. Beer, Chiel, Quinn, Espenschied, and Larsson. A distributed neural network architecture for hexapod robot locomotion.

16. Aleksander, Igor. 1989. *Neural computing architectures.* Cambridge, Mass.: MIT Press.

17. Aleksander, Igor, and Helen Morton. 1991. *An introduction to neural computing.* Chapman and Hall.

18. Wasserman, Philip D. 1989. *Neural computing, theory and practice.* Van Nostrand Reinhold.

19. Mead. *Analog VLSI and neural systems.*

20. Wang, Xiao-Jing, and John Rinzel. 1992. Alternating and synchronous rhythms in reciprocally inhibitory model neurons. *Neural Computation*, Vol. 4, No. 1.

21. Reiss, R. F. A theory and simulation of rhythmic behavior due to reciprocal inhibition in small nerve nets. *Proc. AFIPS Spring Joint Comput. Conf.* Vol. 21: 171–194.

22. Williams, Thelma L. 1992. Phase coupling in simulated chains of coupled oscillators representing the lamprey spinal cord. *Neural Computation*, Vol. 4, No. 4.

23. Heininger and Hendry. Parallel 2 serial controller.

24. Miller, David P., and Marc G. Slack. Global symbolic maps from local navigation. *Proc. of the 9th Int'l Conf. on AI*, Vol. 2: 750–755.

Chapter 14

Cooperative Behaviors of Autonomous Mobile Robots

Susan A. Mengel, James M. Conrad, Lance Hankins, and Roger Moore

INTRODUCTION

The NASA Johnson Space Center is currently exploring the feasibility of space station maintenance and repair via mobile robotic agents. JPL/NASA is working to produce teams of robots to explore planetary surfaces. The Department of the Navy is supporting cooperative robotic research, such as Lynn Parker's work[1] on cooperative heterogeneous mobile robotic behavior in the simultaneous solution of different types of tasks. These and other organizations have realized the benefit of having several simple or sophisticated robots work together to accomplish tasks faster, more efficiently, more successfully, and more cost effectively.

Since robots are known for their high cost, purchasing one sophisticated, expensive robot may not be feasible for many organizations. For example, if the robot fails, the time and cost of repair might not justify its use. If several inexpensive robots could be used in place of an expensive one, work on a task could proceed even if one or two robots fail, thus achieving a high level of fault tolerance. Although repair cost might still be a factor, repair time would become less of a concern with other robots still working. Achieving fault tolerance through robots working together is particularly desirable in mission-critical tasks, such as planet exploration. Other applications of interest include hazardous waste cleanup, aircraft flying, and autonomous vehicles.

Because of the broad range of organizations that have interest in the robotics field, research in cooperative robotics is active and producing results. Currently, research has enabled robots to work together on simple tasks, such as pushing an object or picking up several objects and placing them elsewhere. Even simulations are becoming more complex and can model large teams of homogenous robots carrying out a task.

In order for robots to carry out tasks, they are given varying levels of artificial intelligence (AI), from reactive to behavioral to predictive techniques. Reactive behaviors are local and are produced in response to an external stimulus, such as backing up after bumping into a wall. Behavioral techniques may borrow from the behaviors of biological organisms. For example, applications have been developed in which a herd of robots simulated foraging by gathering items and transporting them to a central location. Predictive behaviors serve to organize robots for a task and include planning to generate the set of tasks for the robots to achieve in a specified order. For example, where task A is to be performed before task B, task A might be performing wall following in a room to determine its dimensions, and task B might be finding an object within the room.

From this description, it is apparent that several disciplines can contribute to the research in cooperative robotics, such as biology, AI, communications, and engineering. The robots and cooperative techniques, therefore, can be used to teach students a wide variety of subjects that naturally arise from their study. Students can learn about mechanical engineering principles from designing robots; they can learn about communication techniques if the robots need to contact each other when working together; they can see how AI techniques help robots work together more efficiently; and they can relate the behaviors of biological organisms to what the robots do to get work done. In many ways, the study of robots can motivate students to look beyond their own discipline to gain a more broad-based education. In the process, they learn cross-disciplinary skills that they would not normally learn in their specific field.

- Students working with control systems go through the life cycle of embedded system development.

- In building a mobile robot and controller board with associated software, students are trained in software-hardware codesign.

- Students learn to apply AI technologies to a complex problem.

Many of these skills are in demand in industry, and students with such skills are highly marketable. Further, the influx of small teams of students trained in cooperative mobile robotics should help support academic, industrial, and government research as well as practical implementations.

Robots, by themselves, have become a favorite high school outreach tool of many university educators, particularly the Stiquito robot or the robots built in the Jones and Flynn book, *Mobile Robots*.[2] Robots have natural appeal to students who like to build things and see them work. Because of their visual appeal, robots can encourage young students to go into areas of high technology, where they can have exciting careers advancing the state of the art in engineering.

Many high school outreach efforts using robots are under way at several institutions, including the NASA Johnson Space Center, Indiana University, and the University of Arkansas.[3,4] These efforts may involve having students assemble robots from instructions to see how they work.

Because of the importance of high school outreach, the authors have been active in using robots to interest students in engineering. Another outreach tool under development at the University of Arkansas illustrates how cooperative behavior can be used with robots to play a game. In this game, called "Hunt the Wumpus," explorers search through a cave to find a sack of gold while avoiding hazards such as bottomless pits and a ferocious beast known as the Wumpus, both of which can kill the explorers. The Wumpus game is used because it is easy to understand, but it is a game that requires some strategy. The Wumpus game can also have explorers cooperatively helping each other find the gold faster, since the entire cave could be searched by one explorer before the gold is found.

The ultimate goal of this tool is to allow students to take the robots they have built and use them to play the game as directed by the computer via radio transmissions. Currently, however, the tool uses a simulation with a graphical user interface depicting Stiquito robots playing the game on an m x n grid so that students can at least see how the robots work together. The interface is written in Borland's Delphi, and a rule-based expert system written in NASA's CLIPS is used to direct the actions of the robots (CLIPS is available through several World Wide Web sites at no cost).

COOPERATIVE ROBOTICS

To give the reader a general perspective of cooperative robotics, along with some of its problems, cooperative robotic systems may be viewed as implementations of intelligent agent architectures or systems. Intelligent systems can consist of one or more intelligent agents designed with distributed, centralized, or hybrid architectures. The architectures for the agents are usually targeted toward specific environments that range from simple to complex.

Historically, uncertain, complex, and noisy environments have caused intelligent agents difficulty in performing their tasks. Such environments can change in unpredictable ways, leaving intelligent agents at a loss as to what to do next. In order to deal with noisy environments, an intelligent agent must be able to adapt itself to new and unusual situations. Recent progress in intelligent agent research has had some success with agent adaptability, particularly in the field of robotics.

Robotic agents have been controlled by reactive, behavioral, hybrid, or planner types of architectures. These architectures are prevalent in the literature and are categorized well by Matarić.[5] Briefly, reactive strategies use table lookup for a set of sensor readings to find the action to take next. They do not store information dynamically (internal state) and, as a result, are inflexible at runtime. Behavior-based approaches do not place a limitation on internal state usage, may involve more computation than just lookup, and can use distributed internal representations and computations to decide what action to take next. Matarić's research into behavior-based systems resulted in 20 homogeneous robots being able to produce the emergent behaviors of flocking and foraging by summing or switching the basic behaviors: safe-wandering, following, aggregation, dispersion, homing, grasping, and dropping. As can be seen, behaviors are an abstraction beyond the sensor or control signals needed to produce the behaviors. Hybrid systems may employ reactive strategies at a low level to provide for the immediate safety of the robot and a planner at a higher level to determine the robot's action sequences. Finally, planners determine all possible action sequences the robot could take for a particular task.

When multiple or distributed agents are involved, some architectures work better than others. Reactive and behavior-based strategies tend to distribute well since they do not necessarily need a centralized intelligence to work, but they might not support complex tasks in noisy environments. One such application involves directing the actions of a robot traveling to several stations to keep bins supplied for manufacturing a product. For example, what happens when a bin is empty, other agents get in the way, bin locations are changed, another agent takes something just before the agent is able to grab it, or something is dropped by an agent and discovered later by another agent? Hybrid and planner systems are able to operate on complex tasks such as this one, but may not be able to do so in a noisy environment. Hybrid and planner systems may also not scale well with an increase in agents or task complexity.

For the interested reader, the following subsections are provided to illustrate a research effort taken for each of the different types of architectures. References to additional literature are also provided.

Reactive Architecture

The University College Salford, U.K., has a team of researchers working on the Behaviour Synthesis Architecture (BSA)[6] implemented both in simulation (up to four robots) and on two actual mobile robots. The architecture is sensor driven, whereby sensor data are immediately converted into control actions by the robot. The robot itself has no map or model of its environment and only responds to external stimuli.

A stimulus generates a motion response along with a utility measure showing the importance of the response. The utility measure is needed for conflict resolution since more than one motion response could be generated from a given stimulus; that is, the utility measure is used to favor one motion response over another. A stimulus or motion response falls into one of four strategy or behavior pattern levels: universe (task-oriented, such as navigation), species (interaction between robots), environment (such as collision avoidance), and self (such as battery level). Behavior patterns are grouped into scripts if they are related.

The researchers implemented this architecture on two robots, Fred and Ginger. These robots together had the task of moving a pallet from one location to another while avoiding obstacles. Even though the robots could not explicitly communicate—they only exhibited reactive behavior—they were fitted with a capture head (instrumented xy table) connected to the pallet via a coned bearing for implicit communication; that is, the displacement of the capture head helped keep the robots working together to move the pallet.

Fred and Ginger were sophisticated robots, even though BSA could be implemented on a conventional 16-bit controller board because it only takes 20 kilobytes. They were driven by three T425 transputers for control and sensing, and a T800 transputer providing run-time storage and debugging. The robots were from Real World Interfaces, Inc., and were based on the B12 platform. They had tuned ultrasonic time-of-flight proximity sensors.

Using Fred and Ginger to push a pallet on a path obstructed with obstacles yielded interesting results. As the collision avoidance utility was turned higher on the robots, they would drop the pallet to avoid an obstacle. As the utility was reduced, they would hang onto the pallet even if it meant bumping into the obstacle. Being able to change the utility function enabled the robots to be more flexible and to handle both fragile and sturdy pallets.

Behavioral Architecture

Matarić successfully implemented cooperative task accomplishment in her *nerd herd*, 20 small homogeneous mobile robots.[7] Taking her inspiration from nature, she borrowed basic behaviors from those found in ants and other organisms, such as *safe-wandering*, *following*, *aggregation*, *dispersion*, and *homing*, and implemented them in her robots. By using the basic behaviors in combination or by switching among them, she found that other, more sophisticated behaviors emerged, such as flocking and foraging.

A behavior is viewed as an abstraction for control, planning, and learning. It is also viewed as a control law for reaching or maintaining a goal. For example, *following* is a behavior whereby the robot takes input from sensors and generates actions to keep moving behind another robot at a fixed distance. *Safe-wandering* helps robots wander without hitting other robots or obstacles. *Aggregation* keeps robots together, separated at most by a specified maximum distance. *Dispersion* separates robots by a minimum distance. *Homing* enables robots to reach a particular location.

By combining the behaviors of homing, dispersion, aggregation, and safe-wandering, robots can be made to flock together to reach a destination. Homing gives them the direction to go, dispersion keeps them from getting too close to each other, aggregation keeps them from getting too far apart, and safe-wandering keeps them from bumping into each other. Foraging, wherein the robots pick up objects and bring them home, can be implemented by switching among the basic behaviors of safe-wandering, dispersion, following, homing, and flocking. First, the robots are dispersed to search for particular objects. If a robot is alone, it uses safe-wandering to keep from bumping into anything during the search. If an object is found, the robot uses homing behavior to bring it to a designated location. If robots with objects encounter each other while going home, they flock together. If robots without objects meet, they follow each other.

A simulation of up to 50 robots was first used to implement Matarić's work, but she also used 20 robots as previously described. Each robot was 12 inches long and had four wheels, one piezoelectric bump sensor on each side and two on the rear, and one two-pronged fork-lift. The forklift had a magnet to hold light, ferrous metal foam-filled pucks; two contact switches on each tip; and six infrared sensors to detect pucks to pick up, hold, and stack. For communication and to distinguish robots from other objects, each robot was fitted with a radio transceiver and, for control, had four Motorola 68HC11 microprocessors: two were used for the radio, one for the operating system, and one for the control system. To help a robot figure out its location, two radio basestations were used for triangulation.

At least 20 trials with the actual robots were used to analyze the basic behavior algorithms to see if they exhibited the criteria of repeatability, stability, robustness, and scalability. Repeatability ensures that the behavior is consistent over different trials, stability checks for oscillations in the behavior, robustness ensures that error and noise do not adversely affect the behavior, and scalability ensures that the behavior is not affected by decreased or increased numbers of robots. The trials successfully showed that the robots could perform the behaviors despite being hampered by position errors and limited range on the infrared sensors. In comparison, a centralized control algorithm was used to implement the behaviors and was faster by a constant factor.

Combining behaviors was successful as well. Flocking turned out to be more efficient than homing since the robots had to maintain a certain distance between themselves and thus moved about in an ordered fashion. Even placing a barrier in front of the robots caused them to split into two groups and rejoin on the other side. Foraging turned out to be slow because agents could not remember where the pucks were, and the pucks changed location because the robots were pushing them around (it took about 15 minutes for two-thirds of the pucks to be picked up). Also, groups did better than single robots in foraging, but had to have more stringent anti-interference monitoring.

Having achieved success with programmed behavior combinations, Matarić turned her attention to having the robots learn combined behaviors, in this case foraging, through unsupervised reinforcement learning (RL). Instead of using states and actions as with typical RL, she used conditions and behaviors, making her RL algorithm more efficient since the number of conditions would be less than the number of states. She also used two types of reinforcement functions: heterogeneous reward, and progress estimator, designed for nondeterministic, noisy, and dynamic environments. The heterogeneous reward function was based on the subgoals required to accomplish a task and rewarded the attainment of the subgoals, while the progress estimators provided a measure of the learner's performance.

Twenty learning trials on foraging were performed using standard Q-Learning, Matarić's RL with the heterogeneous reward, and her RL with both the heterogeneous reward and progress estimator functions. These trials used more reliable robots (four R2 mobile robots) than those in the nerd herd, but the robots had the same functionality. Each trial lasted 15 minutes and RL performed better than standard Q-Learning. RL with the two reinforcement functions fared better than RL with only the heterogeneous reward function.

The heterogeneous reward function for foraging included both positive and negative reinforcement. A robot was positively reinforced for grasping a puck, dropping a puck at home, and waking up at home (an internal day/night clock was added to the robots so they would rest). A robot was negatively reinforced for dropping a puck away from home and waking up away from home. Two progress estimator functions were used: one for keeping agents from getting too close to each other, and the other to show progress made in getting home.

Planner Architecture

Noreils[8] takes the viewpoint that cooperation among heterogeneous robots is facilitated by collaboration (putting robots together to work on a task) along with coordination (synchronizing robotic activities to accomplish the task). He maintains that cooperation implemented through collaboration and coordination can be achieved through the following four steps.

1. *Decomposition.* One robot decomposes the task into subtasks, schedules the subtasks, determines the skills needed, and computes the number of robots required.

2. *Allocation.* The robot puts the subtasks out for bid to available robots and negotiates for subtask accomplishment.

3. *Local planning.* An available robot takes its subtask and plans the sequence of steps necessary to accomplish it. The robot also determines the resources (space, time, physical objects) required to accomplish the task. Since some of the resources may also be required by other robots, conflict resolution may be necessary.

4. *Execution.* The robots begin to work together to accomplish the overall task, and contingencies that occur are handled.

To implement these four steps, Noreils used a three-level planning system.

1. *Functional.* Has modules providing basic processing functions, event detection, and programmable reactivity

2. *Control.* Executes missions or specific tasks, reports information in the event of failures, and organizes the robot's modules to achieve consistent behavior

3. *Planner.* Has high decisional capabilities and is divided into two subsystems: coordination, using coordinated protocols to facilitate task accomplishment; and cooperation, having the responsibility for robot collaboration and local planning

The functional level has three types of modules: sensor/effector, servo-process, and functional. Sensor modules process at increasing levels of abstraction the data received from physical sensors and detect the occurrence of events from given conditions. A sensor module has a sensor surveillance manager attached to it to monitor condition fulfillment and to send a signal to the control level. Effector modules, such as the pilot module for driving the motorized wheels, drive the actuators of the robot. Servo-process modules, such as wall following and obstacle avoidance, have input lines from the sensor modules, output lines to the effector or other modules, and control lines to and from the control level. The control level sends activation or halt signals to the servo-process module and the servo-process module sends failure signals to the control level. Functional modules, such as trajectory planning and robot localization, are activated upon request, may remain permanently activated, and are used to make computations.

The control level has four types of modules: supervisor, executive, diagnostic and recovery, and surveillance manager. The supervisor module receives a plan and controls its execution by sending each plan step (mission) to the executive module to be parsed and fulfilled. The executive module receives missions, manages the modules in the functional level, and detects conflicts, such as multiple access to an effector module. In case of failure, the executive module will call the diagnostic and recovery module, which locates the fault causing the failure and tries to repair the mission or issues a diagnostic on the mission. The surveillance manager module reacts to events in three stages: reflex reaction, action linked to the current mission, and task replanning. If any stage fails to correct a problem, the next stage is executed.

The planner level uses the coordination and cooperation subsystems. The coordination subsystem uses Predicate/Transition nets as defined by Genrich[9] to coordinate a robot's activity with another. The nets allow the coordination of more than one robot and consist of

- Arcs annotated with predicates representing structures with attributes manipulated by modules and propagated to places

- Transitions with formulae representing missions and conditions on attributes of structures propagated through the arcs

- Places that are marked if they have at least one structure in them

Transitions are made if all input places are marked and all conditions are met. The cooperation subsystem is responsible for decomposing the task into subtasks, allocating the subtasks to robots, and local planning. Its decomposition module generates a skeletal task of subtasks, robots and skills required for the task, and constraints. The skeletal task is given to the allocation module, which sends out calls for participation. The negotiation module of a robot receives the call for participation and answers the call. If accepted by the allocation module, the robot generates a skeletal task for itself and inputs this into its local planning module. The local planning module generates a local plan, and detects and solves conflicts. The skeletal task is executed by the execution monitoring module. If a message comes up from the lower levels during execution, the situation assessment module analyzes it, modifies the skeletal task, and puts the skeletal task back into the local planning module, the decomposition module, or the failure report module.

Noreils used two robots to implement his architecture, HILARE 1 and HILARE 1.5. Both robots are fairly sophisticated, with five Intel 80286 and 8086's and a radio link to networked workstations. They both have laser range finders and 16 sensors. HILARE 1.5 has a trinocular stereovision system on a pan platform. Both have two independent drive wheels and free wheels with odometry on the drive wheels. The robots have the functional and control levels implemented on-board, while the planner level is implemented off-board.

HILARE 1.5 was used to supervise HILARE 1 in pushing a box to a wall. To complicate the task, an object was placed in the way of HILARE 1. Both robots were successful in accomplishing the task. HILARE 1.5 performed the overall planning and pushed the object out of the way of HILARE 1 when it seemed the task could not run to completion (HILARE 1 had pushed the box to the object and thought it was finished).

Hybrid Architecture

A three-layered planner, using the RAPs (Reactive Plan Packages) system as the middle layer, is in use at the NASA Johnson Space Center[10] where researchers are working on getting robots to work cooperatively with people in maintaining the space station. The RAPs system was originally developed by James Firby[11] and operates in a goal-based system with internal state. It has three components, each distinct from the other: memory, task agenda, and interpreter. The memory holds the internal state of the robot and contains information about objects currently seen in the area and objects seen in areas the robot has been in the past, thus allowing the robot to recognize objects it has seen before. The task agenda holds those RAPs waiting to run. The interpreter chooses the RAP to execute next.

At a minimum, an RAP consists of an index clause that matches to the current goal to be achieved, a success clause to check that the RAP executed correctly, and a set of methods to achieve success. More than one method is specified so that if one method fails, another can be tried based on the robot's internal state and current information about its external environment. For example, a robot may reach down to pick up an object, but may drop it because its arm is not strong enough to lift it. After one more try, the robot may choose the next method in the RAP, using another arm that is able to lift more weight. An RAP may have other attributes as well, such as timing constraints to ensure execution by a particular time and monitors to check for the occurrence of certain events.

Firby specifically designed his RAP's system to operate in uncertain and dynamic environments, which is why method alternatives were allowed. He further envisioned the RAP's system as being the middle layer of an intelligent architecture, where the top layer is a planner to be used when no RAP exists to accomplish a certain goal and the bottom layer consists of the specific control signals used to operate the robot.

As previously mentioned, the NASA Johnson Space Center took RAPs and placed them into a three-level architecture with a bottom layer of reactive skills, a middle layer of sequencing or RAPs, and a top layer of deliberation or path planning. The researchers at Johnson Space Center have used the architecture successfully with a robotic wheelchair; in implementing following or approaching behavior in robots; with a three-armed EVA helper/retriever robot to carry out maintenance tasks on the space station; and with an errand-runner robot. Their architecture is rather efficient; the bottom layer works in milliseconds, the middle layer works in tenths of seconds, and the upper layer works in seconds to tens of seconds.

OTHER WORK

Chen and Luh[12] have developed movement algorithms for a group of small, identical mobile robots. They identify five subtasks needed to coordinate and control movement of the group.

1. Formation of a geometric pattern
2. Alignment of the orientation of each mobile robot in compliance with the assigned global goal
3. Coordination of the robots in the formation
4. Motion realization of the formation
5. Stability of the formation

They present simulation results demonstrating how 30 mobile robots would iteratively move to form a circle, ensuring that the movements do not result in collisions. They also characterize the computations required for the robots to move in formation.

Researchers from Oxford University have developed a model with formal semantics for multiple vehicle task and motion planning.[13] For example, if several autonomous vehicles are operating in a factory setting, the vehicles will need to communicate to allocate work and coordinate their movements to use the available resources (that is, floor space) efficiently. This model, which has only been simulated, assigns tasks and allocates resources similar to the way tasks are assigned for balancing workload on a parallel computer.

In Japan, Itoh and Anzai[14] are working on a decentralized planner, called Coplanner, implemented on a robot fabricated in their laboratory and simulated with up to 30 robots delivering documents in an office building. Further, a group of European researchers[15] is modeling reactive and distributed predictive behaviors using PROLOG. And Matarić is continuing her work, with two Genghis-II six-legged robots that can push a box to a specified region marked with lights.[16] Other works of interest can be found in the reference list[17,18,19,20,21,22,23,24] and through numerous links on the World Wide Web, such as

- http://ai.eecs.umich.edu/people/durfee/durfee.html
- http://robotics.jpl.nasa.gov/
- http://www.epm.ornl.gov/~parkerle/

HIGH SCHOOL OUTREACH TOOL

From the background on cooperative robotics, one might think that introducing high school students to such a topic is beyond their educational level to understand. Fortunately, high school students can be taught expert systems that use rules to accomplish their work. The rules also can be used to cause cooperative behavior in robots or, in this case, a simulation of several Stiquito robots playing the Wumpus game.

The following sections discuss the Wumpus game, give background on the expert system used to guide the players, and provide figures of the Wumpus game being played. This game is played using three Stiquito robots.

The World of the Wumpus

A simple example of applied AI is the game "Hunt the Wumpus." The world of the Wumpus consists of multiple players wandering through an unexplored cave, attempting to find gold while avoiding such deadly hazards as bottomless pits and a ferocious beast called the Wumpus. The players descend into the cave via a ladder and can explore until they are killed by one of the hazards or until they find the gold and return to the ladder (the winning condition).

The playing area (shown in Figure 14.1) is represented by a square grid of rooms in which the players can move north, south, east, or west (no diagonal moves are allowed). Each room can have several attributes that aid the players in their quest. For example, if a player is adjacent to a square containing a pit, a breeze will be felt. If adjacent to a square containing the Wumpus, the player will detect an awful stench. Should the player ascertain the location of the Wumpus without walking into that square, a single arrow may be shot in its direction, invoking a blood-curdling scream from the beast as it dies. The player scores points during the game under the following conditions:

ACTION	POINTS
Finding gold and safely exiting	+1,000
Each move made	−1
Dying by Wumpus or pit	−10,000

Once the game has been described, one can describe a logical manner in which the players might attempt to obtain the gold safely. Obviously, if a player is in a square and neither a stench nor a breeze can be detected, it can be asserted that all adjacent squares are safe (no hazards). Given multiple breeze or stench squares, a player may sometimes be able to calculate the position of a pit or the Wumpus. The calculations are implemented as inferences in a CLIPS expert system that enables the players to navigate the cave successfully, find the gold, and escape.

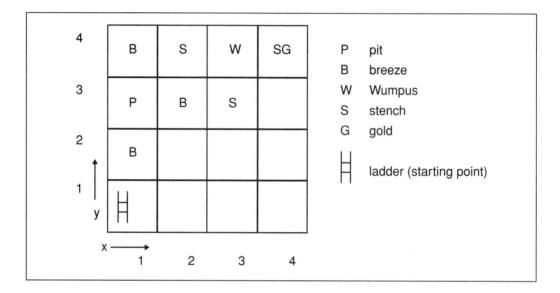

Figure 14.1. Sample playing board.

The Expert Wumpus Player

Within the Wumpus simulation, there are two fundamental types of knowledge: system knowledge and player knowledge. System knowledge is complete with regard to the environment, and it is used by the system to impose environmental or other facts upon the player (for instance, when a player walks into a room with a pit, death results). Player knowledge is a collection of facts known about the player's environment [including such knowledge as the location of a breeze in square (1,2)]. Player knowledge is not complete; it only represents the areas of the environment the player has explored or about which the player has made inferences.

Within CLIPS, both types of knowledge are represented as system-square and player-square fact templates for each grid coordinate within the Wumpus game. They are made up of the following:

system square	player square
x coordinate	x coordinate
y coordinate	y coordinate
Wumpus present	visited
pit present	ladder present
gold present	Wumpus present
stench present	pit present
breeze present	stench present
ladder present	breeze present
	traversion order

At the beginning of the simulation, a rule fires that creates a map composed of an m x n grid having both system squares and player squares. The system squares are initialized (by another rule) to contain various hazards, while the player squares are all initialized to an unknown status (since the player does not know anything about the cave upon entry).

Another set of facts used is the current environmental facts, which represent facts about the room in which the player currently resides. When the player moves into a new room, these facts are updated to match those of the new room.

Several strategies can be used for movement within the Wumpus game. The first and perhaps most straightforward is the traversion count method. As a player moves about in the Wumpus environment, the new squares are numbered as the player encounters them with an ever-increasing traversion count (see Figure 14.2). All squares are numbered as the player leaves them, and the traversion count (initialized to zero) is always equal to the number of moves the player has made at that point. The player never renumbers previously encountered squares.

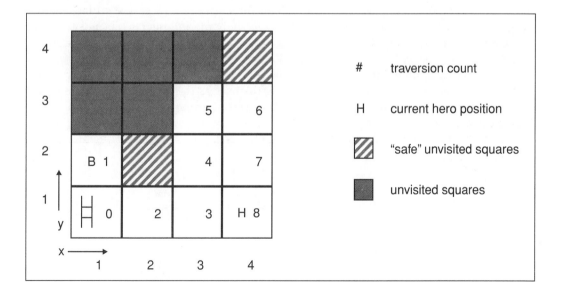

Figure 14.2. Traversion count scenario.

Movement within the traversion count method is governed as follows: If the player is not en route to a specific destination, an arbitrary but safe square (one with no Wumpus or pit) that has not previously been visited is selected as the destination for the player. When the player is en route to a destination that is not adjacent to the current square, the player moves in a manner that *minimizes* the distance in the traversion count from the current square to the traversion count of a square adjacent to the destination square (for the destination square to be marked as safe, it must have at least one adjacent square that has been visited and, therefore, has a traversion count). As shown in Figure 14.2, the player, or hero, is at square (4,1) and decides that the new destination should be (4,4), a previously unvisited safe square. The traversion count movement rule would state that since the count of the only square adjacent to the destination that has been visited [square (4,3)] is 6, the player should move in a manner to enter squares with a traversion count closer to 6 than the current square. This would cause the player to move north into square (4,2), because its traversion count is 7. (The alternative would be to move into square (3,1), which has a traversion count of 3, but this is not as close to the desired count of 6 as the traversion count of the square chosen). Once the player reaches the destination square, the process begins again.

A more efficient (albeit more computationally intensive) navigation technique is the *ripple* algorithm (see Figure 14.3). With this method, when a player has no destination, a high precedence rule fires that causes all safe squares to send out ripples to all adjacent visited squares. These ripples consist of the safe square coordinates and the adjacent visited squares' distance from the safe square (for immediate neighbors of the safe square, this will always be 1). The visited squares that receive the initial ripples in turn send out ripples of their own to their adjacent visited squares (only if a ripple for that square with matching initiating coordinates does not already exist, thus preventing endless loops) with a distance parameter equal to 1 greater than the distance parameter received by the current sender (for the immediate neighbor's ripple to the secondary neighbor, the distance is 2). This rippling continues until no more ripples can be made (as stated, a visited square cannot receive two ripples that were initiated at the same coordinates). It is important to note that different ripples may coexist for the same visited square, provided they have different initiators. Once this process is complete, the player simply chooses the minimum ripple count (distance) from the squares immediately bordering the current position. The coordinates of the ripple initiator on that square will always represent the closest safe square to the player and the trail of descending ripple distance parameters with this corresponding destination coordinate, as the initiator will always represent the shortest safe path to that square. This closest safe square is then set as the destination for the player, and the player continues following the previously asserted ripples with that coordinate as initiator, moving in such a way to follow the descending distance parameters until reaching the destination square. The process is then repeated.

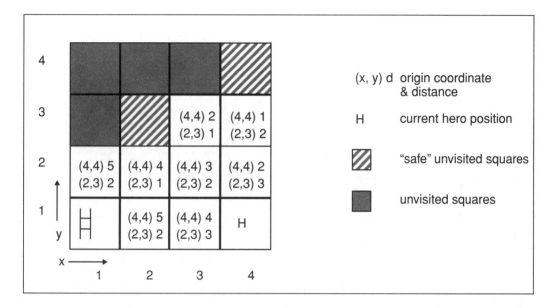

Figure 14.3. Ripple decision scenario.

With either of these navigation methods, should the player happen to find the gold, a higher priority rule will fire that sets the first known player square with a ladder as the new destination (and, if the ripple method is used, initiates a ripple at that coordinate). Once the player reaches the square with the ladder, the player climbs out and wins the game. If at any time no safe, unvisited squares exist, the player will give up and set the destination as the first known player square with a ladder.

Inferences Regarding Unvisited Squares

Inferences are made about squares that have a stench (see Figure 14.4); they help the player know where the Wumpus might be. After identifying where the Wumpus seems to be, the player will attempt to shoot it while positioned at the same x coordinate or the same y coordinate as the Wumpus. If the player kills the Wumpus (a scream is heard), all stench flags are removed from any player squares, which simplifies movement by opening up more safe squares.

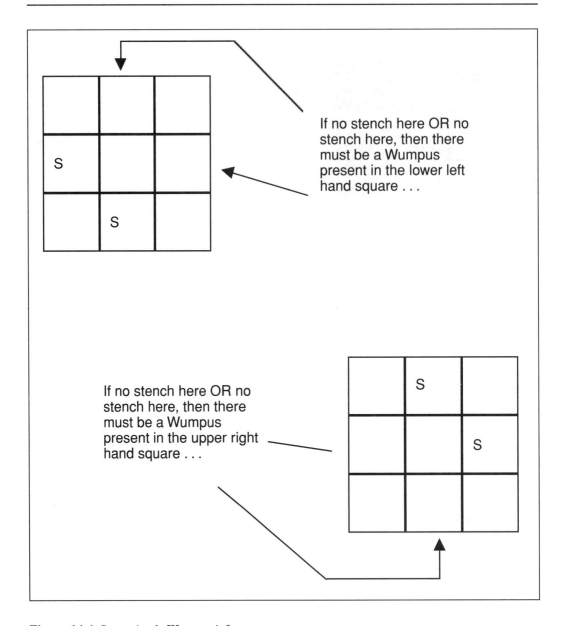

Figure 14.4. Some simple Wumpus inferences.

Using CLIPS to Define the Wumpus Expert System

CLIPS was developed by the Software Technology Branch of the Johnson Space Center, and has proven to be a useful tool in creating expert systems. Although CLIPS was written in C, the structure of the code strongly resembles that of LISP.

CLIPS causes people to think in facts. A known fact list is built up within a CLIPS application as items of truth are continually asserted. For instance, the following commands might be initiated to input facts into the list.

```
(assert (fall_budget   5,000))
(assert (apple   red)
(assert (banana   yellow))
         :
         :
etc.
```

Aside from facts, another major construct within CLIPS is the rule. Rules are made up of a condition-consequence set (upon meeting the appropriate condition, the rule fires or executes and invokes the consequence). In order to create rules, facts (conditions) are specified that must be present for rules to fire, and what should occur upon firing (consequence) must also be specified. An example is the following:

```
(defrule eat_banana_if_good
    (banana_present)
    (banana yellow)
=>
    (assert (banana has_been_eaten))
    (printout t "Banana has been eaten...")
)
```

This simple rule looks to see if the facts (banana_present) and (banana yellow) have been asserted (there is an implicit AND between facts in the "if" clause of a rule when stated in such a manner), and if so, asserts a new fact that now the banana "has been eaten." This is equivalent to saying "if the banana is present and if the banana is yellow, then eat the banana." If the facts that have been asserted are (banana_present) and (banana brown), the rule will not fire. It is important to note that this rule does not remove the presence of the two original facts from the known fact list. In order to remove a fact, a reference to it in the condition clause must be obtained and then it may be retracted. One should not assume that obtaining a reference to the fact precludes the necessity of its presence (the fact must still be present for the rule to fire; rules that contain the statement "?ptr < - (fact1)" still require (fact1) in order to fire). The notation "?name" is used to designate a variable, and the "< -" is used to assign a reference to a fact address.

```
(defrule eat_banana_if_good
    ?present <- (banana_present)
    ?color <- (banana yellow)
=>
    (assert (banana has_been_eaten))
    (printout t "Banana has been eaten...")
    (retract ?present)
    (retract ?color)
)
```

The two retractions remove both "banana_present" and "banana yellow" from the fact list. CLIPS also allows the binding of variables pertaining to portions of asserted facts, and decision making based on the value of those variables. Consider the following example (where a semicolon is used to designate a comment).

```
; assume the following facts have been previously asserted
;
;     (assert (banana_present))
;     (assert (banana  yellow ))
;     (assert (eaten  0))
(defrule eat_banana_if_good
    ?present <- (banana_present)
    ?color <- (banana yellow)
    ; bind the local variable ?num to the value which is paired
with eaten...
    ?count <- (eaten ?num)
=>
    (assert (banana has_been_eaten))
    (printout t "Banana has been eaten...")
    (retract ?present)
    (retract ?color)
    ; retract the fact which designates the old number of bananas
    ; eaten...
    (retract ?count)
    ; assert a new count for the number of bananas eaten...
    (assert (eaten (+ ?num 1)))
)
```

This rule updates the count of bananas eaten as long as an (eaten x) fact exists, where x is some number. Notice CLIPS' prefix notation in the assertion of the new count (the expression A + B} transforms to + A B} in prefix).

As previously mentioned, in developing the expert Wumpus player system, system-square and player-square fact templates are used. A template in CLIPS is roughly comparable to a structure or record in traditional programming languages. Templates allow the consolidation of various and not necessarily identical types (STRING, NUMBER, SYMBOL) of facts concerning an object into corresponding fields (referred to as slots) within the template. Consider the following example.

```
; a simple template to represent the brave hero...
(deftemplate player-facts
    (slot player_number    ; which player (for multiple players)
        (type INTEGER))
    (slot x_position       ; this player's x coordinate
        (type INTEGER))
    (slot y_position       ; this player's y coordinate
        (type INTEGER))
    (slot has_arrow        ; does this player still have an arrow
        (type SYMBOL)
        (default true))
```

```
    (slot has_gold              ; does this player have the gold
        (type SYMBOL)
        (default true))
)


; here is a simplified version of a template representing a grid
; coordinate
; within our simulation, there would be an instance of this
; template for each room within the cave
(deftemplate system-square
    (slot x_position
        (type INTEGER))
    (slot y_position
        (type INTEGER))
    (slot wumpus_present
        (type SYMBOL)
        (default false))
    (slot pit_present
        (type SYMBOL)
        (default false))
    (slot gold_present
        (type SYMBOL)
        (default false))
)
```

These templates might be used to represent a player and a system square within the Wumpus game. Assuming there were a rule or deffacts clause to instantiate appropriate templates for each of the possible rooms into which the player could wander, rules like the following could be written.

```
(defrule grab-gold
    ?pinfo <- ( player-facts (player ?tplayer)  (x_position ?tx)
        (y_yposition ?ty))
    ?sysptr <- (system-square (x ?tx) (y ?ty) (gold_present true))
=>
    (printout t "Player " ?tplayer " grabbed the gold..." crlf)
    (modify ?sysptr (gold_present false) )
    (modify ?pinfo  (has_gold true) )
    (assert (gold_found true))
)
```

Notice that both of the conditions that must be met contain the variables tx and ty. The values of these variables are bound on the first condition only; that is, if the player is at coordinates (5,4), then tx will equal 5 and ty will equal 4, and the second condition becomes

```
    ?sysptr <- (system-square (x 5) (y 4) (gold_present true))
```

Therefore, this rule will only fire if the coordinates of the player correspond to the coordinates of a room that has the gold_present flag set to true [in the example, this rule would only fire if the gold were actually located in room (5,4)]. The modify statement allows changing singular fields within a template. For instance, the statement

```
(modify ?sysptr (gold_present false))
```

simply changes the value of the gold_present slot in this instance of a system-square template to false.

Cooperative Behavior Among Multiple Wumpus Players

As previously mentioned, a penalty of –1 point is incurred for each move made by the player. Suppose the player can only make one move per second, this then is actually a time penalty (–1 point for each second spent in the caverns). Does it then become more point effective to introduce multiple players into the simulation? If special rules are instituted to help initiate cooperative behavior among the players, the answer is yes.

In the Wumpus game, cooperative behavior would ensure that no two players wasted time attempting to reach the same destination square, and that all information, both discovered and inferred, was shared between players. In creating the Wumpus simulation, therefore, the ripple movement algorithm was chosen and augmented with a few other conditions to ensure that, in choosing a new destination for the current player, an unvisited safe square that was already another player's destination (that player is en route to the square, but has not yet reached it) was never allowed to ripple for the current player. While this prevents duplicate destinations and for the most part speeds up the cavern traversal process, it may also, on rare occasion, allow inefficient movement scenarios to occur. In considering Figure 14.5 assume the first player, hero 1 (H1), selects a destination: square (2,3), the closest safe square, requiring three moves to reach. Hero 2 (H2) then selects from the remaining unchosen safe squares, and is forced to choose square (4,4), which will require five moves to reach. The sum of the two players' total required moves is 8. Had hero 2 been allowed to choose first, square (2,3) would have been chosen (requiring two moves to reach), and hero 1 would have been forced to choose square (4,4) (requiring four moves to reach). This results in a total of six moves, which is obviously more efficient. For this reason, the movement strategy could stand tempering with several other logic rules.

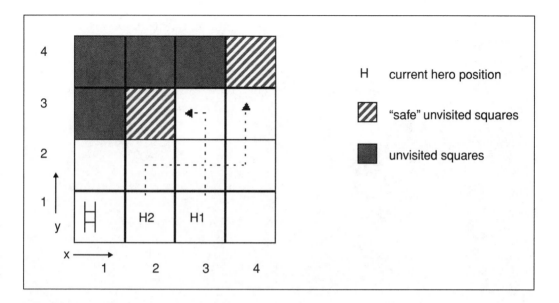

Figure 14.5. Possible inefficiency.

PLAYING THE WUMPUS GAME

A graphical user interface was written in Borland's Delphi and compiled to a Dynamic Link Library (DLL). CLIPS was recompiled to call the interface routines through the DLL to update the screen as the Wumpus game is played. The interface is illustrated in Figures 14.6 through 14.12, which show the progression of the three Stiquito robots as they play the game and eventually wind up with the gold in Figure 14.12. The screens enable students to visualize how robots can work together to accomplish a simple task.

Figure 14.6 shows that the robots have climbed down the ladder and have started to search the cave. Figure 14.7 shows that the robots have found the southwest corner of the cave to be blocked by breezes (indicating pits) and are starting to venture eastward. Figure 14.8 shows the robots having searched the south portion of the cave. Figure 14.9 shows the progression of the robots avoiding the pits as they approach the north and central portions of the cave. In Figure 14.10, the robots have finally encountered a stench square, so the location of the Wumpus is about to be discovered. In Figure 14.11, the robots have killed the Wumpus and are still seeking the gold. Finally, in Figure 14.12, the robots have found the gold located just north of the Wumpus.

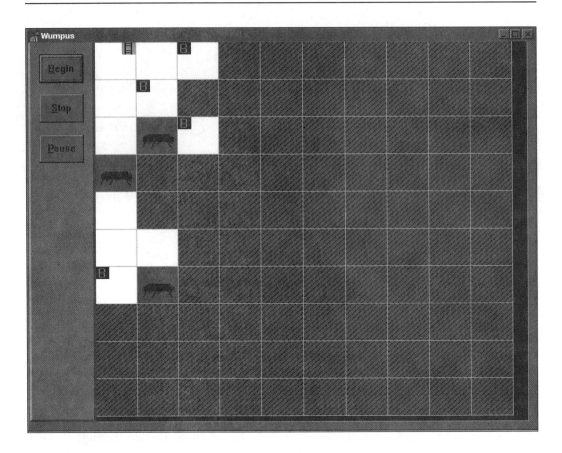

Figure 14.6. Starting out.

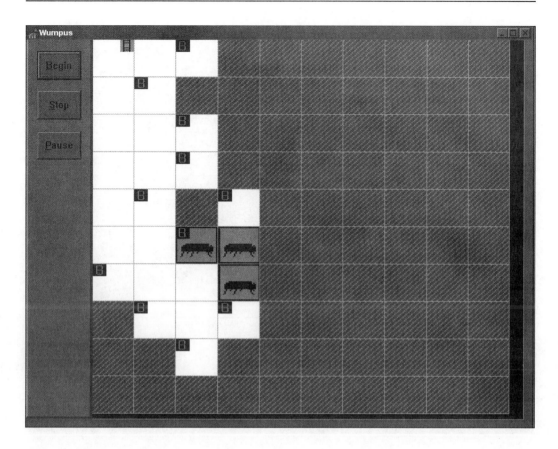

Figure 14.7. Avoiding the pits in the SW corner.

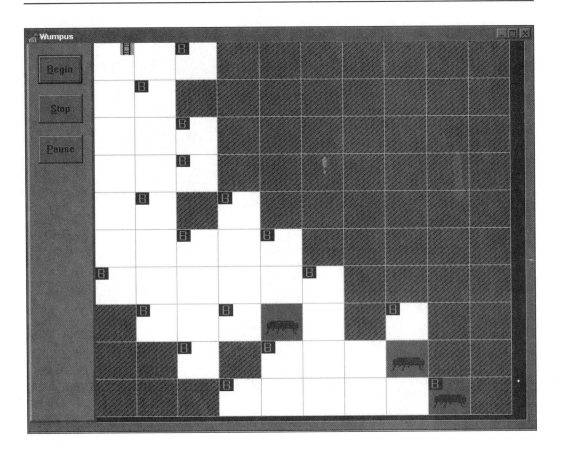

Figure 14.8. Exploring the south part of the cave.

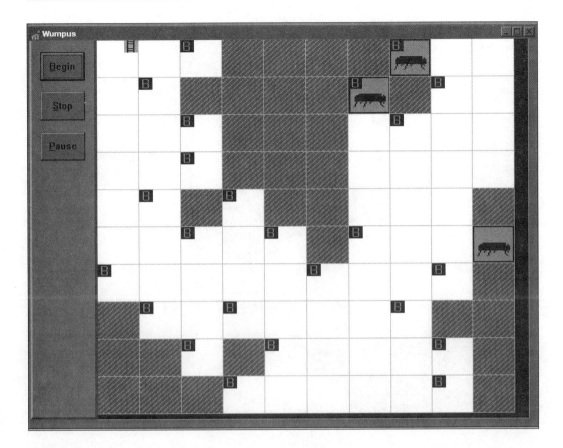

Figure 14.9. Moving toward the northern and central portions of the cave.

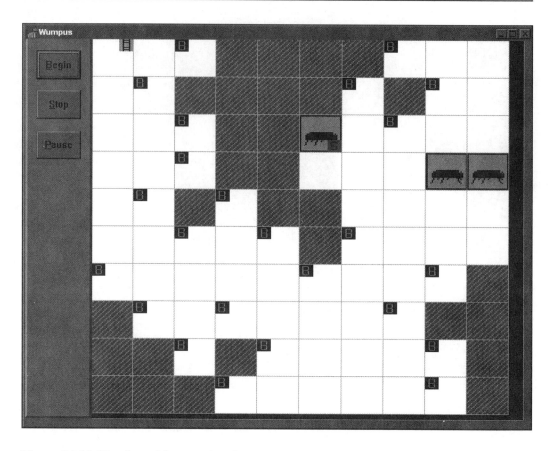

Figure 14.10. First "stench" square found.

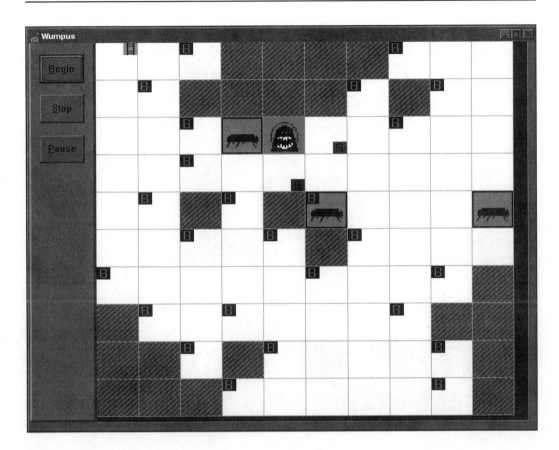

Figure 14.11. The Wumpus shot dead.

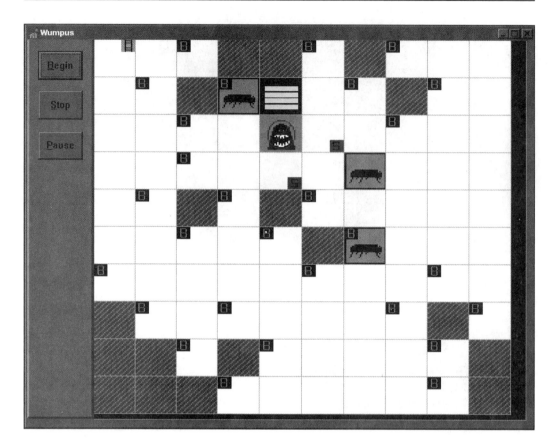

Figure 14.12. The gold found.

FUTURE WORK

Visual demonstrations enable students to see how a problem can be solved. Once they see what is to be done, they are usually more receptive to, and are often curious about, the underlying mechanisms that are used to solve the problem. If students can see robots working cooperatively together, some might have their interest sparked enough to go into engineering and learn how they can make robots work together themselves. Robots can inspire students in different ways, since robots are interdisciplinary, representing the mechanical, computer, industrial, electrical engineering, and other disciplines.

Making engineering real to high school students is part of the motivation for the high school outreach tool. Although no single tool can be used to bring all high school students into the fold of engineering, never using the tool at all may reinforce the myth that engineering is cold and a hard discipline to learn. If students can be made to see what engineering is like, there is always the chance that some students who might not have considered engineering will at least think about it as a career option.

To encourage students to choose engineering as a career, the high school outreach tool will be modified in two ways in the short term. First, the tool will have a multimedia tutorial added on expert systems. The tutorial will be written in SIMPLE, developed by Marion Hagler and Bill Marcy at Texas Tech University and available free to educators. After completing the tutorial, students will construct a small expert system using CLIPS. Second, the Wumpus game will be played with actual robots, such as Stiquito or LEGO, playing on a grid and receiving commands via a radio link from a computer running the game. The graphical user interface will be used to show the progress of the robots. The robots may even be built by the students with a radio transceiver to be attached later when the game is played.

The tutorial can also be expanded in the long term to include more material on cooperative robotics and more simulations of cooperative behavior. Additional experiments with actual robots may be added to illustrate the concepts in the tutorial. The tutorial would then be useful for training college students quickly in the field of cooperative robotics to get them ready to do research or practical applications with robots. The tutorial should at least contain one example of each of the four basic types of intelligent architectures: reactive, behavioral, planner, and hybrid. Students would then see for themselves how the architectures work.

Besides the tutorial, the simulation might be expanded to run on a parallel computer, such as a transputer. Since the robots are being moved by one processor (the PC), all but one remain still while another robot is computing the next safe square destination. Running each robot on its own processor would enable it to keep moving even if other robots were figuring out where to go next. The Wumpus could also be automated, move undetected about the grid, and be more of a danger since the robots would not be able to locate it safely.

The simulation can also be modified to plug in different intelligent agent techniques to see how each affects the playing of the game. In fact, the game can be made more complicated to see how each technique responds to increasing levels of difficulty. For example, the robots could be made to operate by using behavioral techniques and made to search for the gold in more than one square. Reactive techniques could be used on the original game to see how well the robots avoid hazards. Predictive techniques could be used to plan a path through the cave and change the plan whenever hazards are found.

SUMMARY

Cooperative robotic techniques provide a viable alternative to using one, expensive, monolithic robot that can fail, causing a task to be aborted. In its place, several smaller robots can be used to accomplish the task by using reactive, behavioral, or predictive techniques. Reactive techniques simply respond to external stimuli and are not computationally intensive. Behavioral techniques distribute well and allow robots to take on the behaviors of biological counterparts, such as ants. Reactive techniques are computationally intensive and plan the task before the robot embarks upon it.

Cooperative behaviors can be shown using the Wumpus game. Although not a complex game in itself, it does take strategy to avoid the dangers of the cave (and being killed). The players can be controlled through intelligent agent techniques to explore the cave and find the gold. If the intelligent agent technique used is rule based, then high school students can be shown how cooperative behaviors work with robots. If more sophisticated intelligent agent techniques are used, college students can use self-discovery learning to see how each of the techniques affects how the game is played and the level of difficulty the game can have.

ACKNOWLEDGMENTS

This work is supported by the Arkansas Space Grant Consortium and the University of Arkansas.

Many thanks to the students who have helped in this effort. They include Ravi Tandon, Jay Parikh, Jeff Bateman, Robert Turner, and Cameron Porter.

All terms known to be trademarks or registered trademarks are capitalized.

REFERENCES

1. Parker, L. 1994. An experiment in mobile robotic cooperation. *Robotics for challenging environments.* New York, N.Y.: Am. Soc. of Civil Engineers, 131–139.

2. Jones, J.J., and A.M. Flynn. 1993. *Mobile robots: Inspiration to implementation.* Wellesley, Mass.: A.K. Peters.

3. Conrad, J.M., and J. Mills. 1994. Inexpensive technology lab exercises for grades 6–9. *Frontiers in Education Proc.* Piscataway, N.J.: IEEE Computer Soc. Press, 218–222.

4. Conrad, J.M. 1994. Introduction to engineering concepts for middle, junior high, and high school teachers. *Frontiers in Education Proc.* Piscataway, N.J.: IEEE Computer Soc. Press, 250–252.

5. Matarić, M. 1994. Interaction and intelligent behavior. Technical Report AITR-1495. Massachusetts Institute of Technology Artificial Intelligence Laboratory.

6. Eustace, D., R.S. Aylatt, and J.O. Gray. 1994. A behaviour synthesis architecture for co-operant mobile robot control. *Control 94.* London, U.K.: IEE, 549–554.

7. Matarić. Interaction and intelligent behavior.

8. Noreils, F.R. 1993. Toward a robot architecture integrating cooperation between mobile robots: Application to indoor environment. *Int'l J. of Robotics Research*, Vol. 12, No. 1: 79–98.

9. Genrich, H. 1986. Predicate/transition nets. *In Petri nets: Central models and their properties—Lecture notes in computer science.* New York, N.Y.: Springer-Verlag, 319–327.

10. Bonasso, R.P., D. Kortenkamp, D.P. Miller, and M. Slack. 1995. Experiences with an architecture for intelligent, reactive agents. 1995 IJCAI Workshop on Agent Theories, Architectures, and Languages.

11. Firby, R.J. 1989. Adaptive execution in complex dynamic worlds. Technical Report YALEU/CSD/RR/#672, Computer Science, Yale Univ.

12. Chen, Q., and J.Y.S. Luh. 1994. Coordination and control of a group of small mobile robots. *Proc. 1994 IEEE Int'l Conf. Robotics and Automation, vol. III.* Los Alamitos, Calif.: IEEE Computer Soc. Press, 2,315–2,320.

13. Rugg-Gunn, N., and S. Cameron. 1994. A formal semantics for multiple vehicle task and motion planning. *Proc. 1994 IEEE Int'l Conf. Robotics and Automation, vol. III.* Los Alamitos, Calif.: IEEE Computer Soc. Press, 2,464–2,469.

14. Itoh, Y., and Y. Anzai. 1993. Cooperative task planning for autonomous mobile robots. *Systems and Computers in Japan*, Vol. 24, No. 14: 95–107.

15. Chaudron, L., J. Erceau, and B. Trousse. 1993. Co-operative decisions and actions in multi-agent worlds. *Proc. IEEE Int'l Conf. Systems, Man, and Cybernetics, vol. III.* Piscataway, N.J.: IEEE, 626–629.

16. Matarić, M., M. Nilsson, and K.T. Simsarian. 1995. Cooperative multi-robot box-pushing. *Proc. IROS 1995.*

17. Donald, B.R., J. Jennings, and D. Rus. 1994. Analyzing teams of cooperating robots. *Proc. 1994 IEEE Int'l Conf. Robotics and Automation, vol. III.* Los Alamitos, Calif.: IEEE Computer Soc. Press, 1,896–1,903.

18. Eustace, D., R.S. Aylatt, and J.O. Gray. 1994. Combining predictive and reactive control strategies in multi-agent systems. *Control 94.* London, U.K.: IEE, 989–993.

19. Morignot, P., and B. Hayes-Roth. 1995. Adaptable motivational profiles for autonomous agents. Technical Report KSL 95-01, Knowledge Systems Laboratory, Computer Science, Stanford Univ.

20. Ishida, Y., H. Asama, S. Tomita, K. Ozaki, A. Matsumoto, and I. Endo. 1994. Functional complement by cooperation of multiple autonomous robots. *Proc. 1994 IEEE Int'l Conf. Robotics and Automation, vol. III.* Los Alamitos, Calif.: IEEE Computer Soc. Press: 2,476–2,481.

21. Johnson, P.J. 1994. Cooperative control of autonomous mobile robot collectives in payload transportation. Master's thesis, Virginia Polytechnic Instit. and State Univ.

22. Ram, A., R. Arkin, G. Boone, and M. Pearce. 1994. Using genetic algorithms to learn reactive control parameters for autonomous robotic navigation. *Adaptive Behavior,* Vol. 2, No. 3: 277–304.

23. Rush, J.R., A.P. Fraser, and D.P. Barnes. 1994. Evolving co-operation in autonomous robotic systems. *Control 94,* London, U.K.: IEE, 995–999.

24. Wolkomir, R. 1991. Working the bugs out of a new breed of "insect" robots. *Smithsonian,* Vol. 22, No. 3: 65–73.

Chapter 15

The Simulation of a Six-Legged Autonomous Robot Guided by Vision

Paulo W.C. Maciel

INTRODUCTION

In the past 20 years many simple autonomous robots have been built at prestigious institutions such as the Massachusetts Institute of Technology and Carnegie Mellon University. Many of these mechanical creatures are guided by several different kinds of sensors, from infrared proximity sensors to sonar and, ultimately, vision systems.

Problems with these kinds of sensors range from the variation in sensitivity between two infrared sensors and the angle of sonar beams, to properties of objects such as size and surface albedo. With respect to vision the problem is even more difficult. A task such as color discrimination, which may be trivial for a human being, is simply an open research problem! The problems of vision systems become even worse when trying to analyze a sequence of images taken from a robot with a head-mounted camera, since these images tend to be noisy.

The objective of this work is the simulation using computer graphics of a six-legged autonomous robot (Stiquito-like) that wanders inside a maze, solely guided by what it sees through a built-in camera located in its forehead.

Although computer simulation will not find solutions to all of the problems faced by engineers when designing and constructing real autonomous robots, a good simulation should enable the study of aspects of the vision problem, which can then be tested and improved in real machines.

The first section examines the navigational problem and shows how the robot, based on what it sees, can find its way through a simple maze without bumping into the maze walls. The simulation section then describes how the simulation was created. The section that follows describes how vision information was extracted from the image seen by the robot. It also lists the vision algorithms used to determine when the robot is too close to a wall and must stop walking, how it recognizes a corner, when it should stop turning, and when it needs to adjust its course while heading toward a wall. Finally, conclusions are drawn about the experiment, and directions are given for further experiments, including some of the possible applications of a simulation of this kind allowing for more realistic scenarios.

DEFINITION OF THE PROBLEM

The problem statement of this work is as follows:

Given a six-legged robot capable of moving forward and making right and left turns and a maze composed of a set of walls organized in such a way that each adjacent wall forms a 90-degree angle with the other, how can the robot be made to travel as far as it can inside the maze without bumping into the walls, relying only on what it sees from a head-mounted camera.

In this simple scenario there are only a few possible kinds of images that the robot vision system should be able to detect:

- At a point where two maze walls and the floor meet, there is a fork, a Y-like junction.

- Inside a corridor, the image seen is a set of four lines in perspective, going toward a square that corresponds to the wall at the end of the corridor.

- At the point of a corridor where one of the maze walls ends (either left or right) and meets the floor, there is an L-like junction.

- When the robot faces a wall, the image seen is of two horizontal lines.

Based on these simple scenes inside the maze, which are made even simpler by using an edge detector,[2] the rules used to guide the robot are now described.

Determining the Distance to a Wall

While the robot walks toward a wall, the processed image (the one that contains an edge-detected version of the camera view) shows two horizontal lines that move apart: the bottom one moving down and the top one moving up. At each step this image is analyzed and the distance from the bottom of the window that contains the processed image to the bottom horizontal line is measured. When this distance falls below a preestablished threshold the robot knows that it is too close to the wall and it needs to stop.

The robot's field of view is very sensitive to the distance at which it decides to stop before hitting a wall. If the robot stops too close to the wall it will only see the wall and nothing else; it will not be able to see a corner, either to the right or the left. If, however, it stops too far from the wall, it might see a scene too complex—formed by the way corridors are organized in the maze—and be unable to understand what it sees. The behavior of the robot becomes too constrained by the threshold distance established in the program.

This problem is overcome by allowing the robot to look to its right and its left and search for corners whenever it comes too close to a maze wall.

Deciding Which Way to Turn

After stopping in front of a wall and looking to the left and right, there are four possible actions that the robot can take.

1. If it sees a corner to its left and no corner to its right, the only way to go is to the right; the robot initiates a turn to the right.

2. If it sees a corner to its right and no corner to its left, it initiates a turn to the left.

3. If there are no corners either to the left or the right, both directions are equally good; the robot makes a random decision and starts turning toward the chosen direction.

4. If there is one corner to the left and one to the right, the robot has reached a dead end and, since it does not back up, it stops.

Deciding When to Stop Turning

Having initiated a turn, the robot monitors what it sees and decides when to stop turning. As the robot starts turning, the two parallel horizontal lines that were seen when the robot decided to stop (because it was too close to a wall) become slanted and converge toward a point in the image. Seeing these two slanted lines, the robot keeps turning.

Eventually the robot will begin to see the end of the corridor; when it is oriented in a direction perpendicular to the wall at the end of the corridor, it stops.

Monitoring the turn is done by selecting the midpoint of the processed image, which corresponds to where the camera is located (since this image is constrained to show the center portion of the camera image), and drawing a vertical line that intersects the wall's edge.

When a 90-degree angle is formed by this vertical line and the line that determines the edge between the wall and the floor, the robot stops turning and starts walking again.

Deciding When and How to Adjust Course

After making a turn, the horizontal line corresponding to where the wall at the end of the corridor meets the floor appears perpendicular to the vertical line from the center of the window. As the robot walks, however, the horizontal line becomes jagged because of an initial imperfect alignment of the robot with the corridor. That is, when it starts walking again, the robot is not perfectly perpendicular to the wall at the end of the corridor. The edge becomes discontinuous either to the right or to the left by two or more pixels, depending on the degree of misalignment between the robot's heading and the wall.

Because the robot does not check its distance from the right and left walls while walking down a corridor, if it is not heading perpendicular to the end wall (and depending on the size of the corridor) it will eventually hit the lateral wall. A course correction is thus required.

Course correction is performed with each of the robot's forward steps. At each step forward, the vertical line from the center of the window must form a T-junction with the bottom edge of the corridor's end wall. The robot monitors the edge forming the top of the T: Whenever it becomes jagged to the right, the robot stops and initiates a turn to the left until the horizontal line becomes continuous again. If the discontinuity is to the left, the robot initiates a turn to the right.

After the course-correcting turns, the horizontal line is continuous and the robot is again aligned with the corridor. It then resumes its normal course forward until another course correction is required or the corridor ends.

THE SIMULATION

This simulation was initially performed on a PersonalIris Silicon Graphics workstation running Iris using the GL graphics library[3] functions, which provide z-buffering, perspective drawing, and the ability to read the contents of the frame buffer and write pixels to a window. Later, a port was made to a Silicon Graphics Crimson with a RealityEngine graphics board for improved performance.

The animation is accomplished by a loop in which, whenever something changes in the scene such as a movement of one of the robot's legs, the whole scene is completely redrawn. The redraw is done to a "backbuffer" frame buffer, which is then swapped with "frontbuffer" to be displayed. The next scene is again drawn to the backbuffer and the animation continues. The robot's gait is modeled after a real insect.[4]

The robot is drawn by putting together 13 stretched cubes and one sphere to simulate the robot's eyes, and the floor is a loop that draws squares on the xz plane. The maze is drawn by a series of rotations and translations from the origin of the coordinate system of walls of three different sizes (small, medium, and large) that are read from a file. The following line specifies one of the walls of the maze.

5 m 90 −3 0 3 0 0 200

This describes wall number 5 as being a medium-sized wall rotated 90 degrees from its default position, which is then translated to coordinates (−3,0,3). The last three numbers represent the RGB wall color (blue in this example). The walls (s, m, and l) are defined to be at the origin and are parallel to the xy plane. Using this description, different maze files allow different simulation environments.

The simulation provides three different windows as shown in Figure 15.1. The large window displays the maze, the floor, and the robot walking on it. The second window (camera view) is drawn from the position of the robot's eye; it displays what the robot is actually "seeing." The third window contains a processed version of part of the camera view. This is the central part of the camera view image when the robot is heading toward a wall, and it is the right or left portion of the camera view when the robot is stopped in front of a wall and is looking right or left. This processed image is a black-and-white version of the camera view after the corresponding part of this image is passed through a derivative operator, which extracts the edges of the scene.

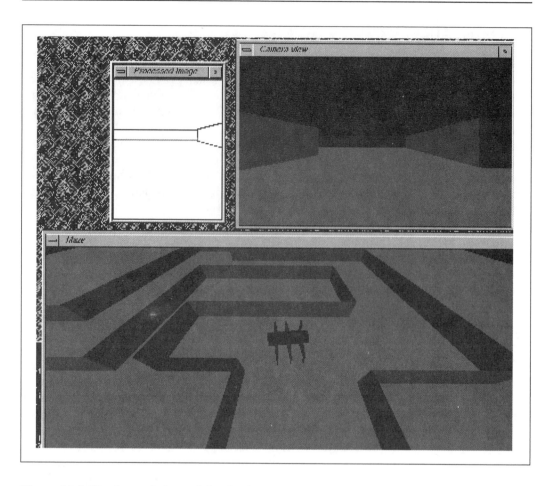

Figure 15.1. The three windows of the simulation.

THE VISION ALGORITHMS

Before the vision algorithms can be applied, the edges on the camera view window need to be detected and drawn onto the processed image window. These edges are obtained in two steps.

1. The vertical and horizontal discrete derivatives of the image are obtained and converted to contain only black and white pixels. The derivative is obtained by making pixel(i,j) = pixel$(i + 1,j)$ – pixel$(i – 1,j)$ for the horizontal derivative and pixel (i,j) = pixel$(i, j + 1)$ – pixel$(i, j – 1)$ for the vertical derivative.
2. These two derivatives are then "ored" to obtain the final edge-detected image.

This edge detector was chosen because of the simplicity of the scenes and its implementation. A more elaborate edge detector that computes the zero crossings of the Laplacian[5] of the image could have been used to achieve the same results.

For every step forward or turn performed by the robot, a portion of the camera view is read into an array in memory that is then processed by the algorithm detailed above to become the processed image in Figure 15.1. This black-and-white edge-detected image (the edges are black lines on a white background) is then ready for recognition by the vision algorithms.

To better visualize the workings of the vision algorithms described in the following sections, this image is written back to the processed image window, and as the algorithms progress the appropriate pixels are colored red.

Four algorithms are used to do the following:

1. Determine when the robot is too close to a wall and needs to stop walking.

2. Determine when the robot should start turning and in which direction it needs to turn.

3. Determine the need for a course correction so that the robot is again aligned with the corridor.

4. Determine when the robot is aligned with the corridor and can stop turning.

It should be pointed out that since these algorithms work by checking black pixels on a white background they are sensitive to aliasing,[6] although they can be made more robust if needed.

Algorithm to Detect Distance to a Wall

The distance between the robot's forehead and a wall is measured in terms of the number of white pixels in a vertical line starting at the midpoint of the bottom of the processed image to the horizontal line that represents the bottom of the wall. If this number is less than a preestablished threshold the robot has to stop walking, otherwise it continues until it reaches the threshold. The getStartingPixel algorithm that follows is used to measure this distance and returns both the position of the last black pixel and the number of vertical black pixels in the vertical line.

Input: The array containing the processed image.

Output: Number of black pixels in the vertical line, last black pixel in the vertical line.

```
int getStartingPixel( Pixel *current )
{
    *current = black pixel at the midpoint of the bottom of the processed image array.

    /* Process vertical line. */
    while( not end of image array )
        if ( peekNextPixel('up') == WHITE ) {
            *current = getNextPixel('up');
            count++;
        }
        else break;
    return count;  /* Number of pixels in the vertical line. */
}
```

Algorithm to Detect When to Start a Turn

This algorithm starts at the pixel returned by getStartingPixel and uses the hasFork algorithm to follow the pixels at each side of the initial pixel until it finds an L, a fork, or the indices of the processed image array fall off the array. Based on where the fork is, the robot decides whether to initiate a turn in the opposite direction.

Pseudocode for the hasFork algorithm is as follows:

Input: The array containing the processed image, the current pixel, and the side being checked for (either left or right).

Output: TRUE/FALSE - found (or not) a fork.

```
Boolean hasFork( Pixel current, Side side )
{
  /* Process horizontal line. */
  do {
    if ( There is a vertical black line starting at the current pixel ) {

      if ( There are no adjacent black pixels to the 'side' of the current pixel )
        return( FALSE );       /* It is an L. */
      else
        return( TRUE );        /* It is a fork. */


    current = getNextPixel( side ).

  } while( not end of image array )

  return( FALSE );
}
```

The algorithm that determines the need to start a turn is as follows:

Input: The array containing the processed image and the current pixel.

Output: TRUE/FALSE - need (or not) to make a turn and the side of turn.

```
Boolean startToTurn( Pixel current, Side *side )
{
  Boolean hasLeftFork = hasFork( current, 'left' );
  Boolean hasRightFork = hasFork( current, 'right' );
```

```
if ( hasRightFork && hasLeftFork )
    return FALSE;           /* Robot reached a dead end. */

if ( hasRightFork && !hasLeftFork )
    *side = 'left';         /* Start turning left. */

if ( !hasRightFork && hasLeftFork )
    *side = 'right';        /* Start turning right. */

if ( !hasRightFork && !hasLeftFork )
    *side = getRandomSide();   /* Start turning to an arbitrary side. */

return TRUE;
}
```

Algorithm to Detect When to Stop Turning

Starting at the current pixel (obtained by getStartingPixel) this algorithm follows the pixels to both the right and left until it finds a discontinuity, an L junction, or the array index falls off the image array. If it finds an L or the array index falls off the image array on both the right and left, the line is considered to be horizontal. The pseudocode for this algorithm is as follows:

Input: The array containing the processed image.

Output: TRUE - found a straight line; i.e., can stop turning. FALSE - no straight line was found; i.e., continue to turn.

```
Boolean stopTurning()
{
    Pixel current = getStartingPixel();

    if ( isContinuous( current, 'right' ) &&
         isContinuous( current, 'left' ) )
        return TRUE;

    return( FALSE );
}
```

Algorithm to Detect Need for Course Correction

If there is either left or right discontinuity in the horizontal line representing the bottom edge of a wall toward which the robot is walking, the robot can correct its course by making a turn in the opposite direction.

This algorithm makes use of the isContinuous algorithm, which determines whether a line is continuous to the left or right of the current pixel (determined by getStartingPixel( )).

Input: The array containing the processed image, the current pixel, and the side to be checked.

Output: TRUE/FALSE - found (or not) a straight line to 'side' of the current pixel.

```
Boolean isContinuous( Pixel current, int side )
{
  do {

    if ( There are no black pixels "strictly" to 'side' of current pixel )
      return( FALSE );   /* found a discontinuity */


    /* Check for a vertical line starting at the current pixel. */
    if ( There is a vertical line at the current pixel )
      return TRUE;   /* Found an 'L' */


    current = getNextPixel( side ).

  } while( not end of image array )

  return TRUE;
}
```

The course correction algorithm is as follows:

Input: The array containing the processed image, the current pixel.

Output: TRUE/FALSE - found (or not) a continuous line and the side to which the robot should turn to correct course.

```
Boolean needCourseCorrection( Pixel current, Side *side )
{
  Boolean rightContinuous = isContinuous( current, 'right' );
  Boolean leftContinuous = isContinuous( current, 'left' );
```

```
if ( rightContinuous && leftContinuous )
    return FALSE;              /* No course correction is needed. */


if ( rightContinuous && !leftContinuous )
    *side = 'right';          /* Start turning left. */


if ( !rightContinuous && leftContinuous )
    *side = 'left';           /* Start turning left. */


return TRUE;
}
```

CONCLUSIONS AND EXTENSIONS

This chapter has described the simulation of a Stiquito-like walking robot that finds its way through a simple maze by being able to interpret what it sees and make navigational decisions based on that interpretation.

The simulation perfectly shows that the robot's judgments were indeed correct; when the experiment was run with a 22-wall maze it was able to walk, make correct turns, and adjust its course. Nevertheless, this experiment is still limited with respect to the orientation of the maze walls (as it is, two adjacent walls must form a 90-degree angle) and to the realism of the scenes the robot sees.

In a more realistic experiment, the robot would wander inside a maze (possibly an office or an industrial plant) identifying landmarks, which could be walls with certain colors, textures, or drawings, or three-dimensional objects along the way, and build an internal representation of the terrain covered together with the location of the seen objects. Ultimately, what is desired is to be able to tell the robot things like, "Go pick up the metallic bar in front of the wooden wall." Clearly, this is not a simple task.

Some of the features that would be needed to accomplish such a task include the following:

- A representation of the seen object, possibly using techniques that recover shape from shading
- Matching techniques that would interpret the acquired information as being similar to generalized representations of prestored known objects in the robot's memory
- Techniques such as the Hough transform to identify patterns on walls or the outline of objects
- Techniques such as stereopsis to determine the length of a corridor

The possibilities are numerous. Although simulations like the one described in this chapter are not likely to provide solutions to real physical problems, they provide a platform where many ideas can be tested.

REFERENCES

1. Brooks, R.A. The role of learning in autonomous robots. Technical report, MIT AI Lab.
2. Ballard, D.H., and C.M. Brown. 1982. *Computer vision.* Prentice Hall.
3. Silicon Graphics Inc. *Graphics library programming guide.* 1990.
4. McKenna, M. 1990. Dynamic simulation of autonomous legged motion. *Proceedings of SIGGRAPH.*
5. Ballard. *Computer vision.*
6. Silicon Graphics.

Chapter 16

The Future for Nitinol-Propelled Walking Robots

Mark W. Tilden

> *The present would be full of all possible futures, if the past had not already projected a pattern upon it.*
>
> —André Gide

When you hear about it for the first time, nitinol is one of those materials you think they made up during an episode of "Star Trek" or something. When it was first announced that a *robot muscle* had been invented and that it was usable, strong, and lightweight, the resounding cry was, "Boy are we gonna see some cool robots now!" As with most things however, reality tempered imagination, and like high-temperature superconductors, it turns out there are places where you can use nitinol and places you cannot (fortunately, these places are a bit more accessible than the shady side of Pluto). With this change in perspective, these limitations can tell us where and how nitinol can be used effectively, rather than where it has a problem.

Right now the future of nitinol seems to depend on getting some new and exotic applications. And since shower-temperature regulators aren't very glamorous, that means interesting robot mechanisms—and that means legs. It doesn't mean *just* legs, but any form of new, complete machine based on the needs of the design rather than biologically inspired ideals (and if these happen to include legs, so much the better). Robots, especially nitinol machines, are based on different materials than nature has used, and nitinol robots must take advantage of this difference, not be limited by it. Therefore, if you make a machine, make it for itself, not because others think it should "look like a bug." When building a robot, the weakest design feature can turn a good idea into a useless pile of junk. It is best to think about every facet of a proposed design, get the parts to help you decide what will be what, and then build for optimal function and reliability.

The constraints in an operable independent robot design are knowing and integrating how is it built, how is it controlled, and how is it powered. Nitinol breaks under high loads, so heavy batteries are out unless you put in a lot of legs. Computers are heavy, so you should either get some chipboard computers or think about a hardwired controller, something that can take a lot of current. Solar cells and high-storage capacitors are easy to use and obtain, and together they weigh significantly less than most batteries. Furthermore, if the device is solar powered, it won't need a power switch, recharger plugs, mounts, and so on. The complexity of the device can be minimal, especially if you consider using simple coupled oscillators as the leg controller. No program to upload means small size, weight, and cost. In all, it is a minimal-fuss machine.

Keep it simple, but try to make it complete. A device that can move, direct, and power itself without tethers is more creature than machine. And if we consider a particular nitinol creature design as a first attempt at robot evolution using humans as the reproductive medium, we can accept the first designs as crude, bizarre things (at first nature wasn't pretty either). Of course, this does require a commitment that more than one machine be built.

Nature has the advantage of infinite experimentation space, but fortunately human designers can stockpile direct experience so that subsequent robot generations can benefit from previous attempts. With this "robogenetic" technique, it is hoped that things might move a little faster than a billion years between upgrades.

The advantage to this type of research is that machines can be studied over long periods (months) and in situations that currently defy simulation (such as on shifting piles of sand and rocks). Extending the keep-it-simple design rule has resulted in machine evolution from 2-transistor rovers to 24-transistor walkers, tumblers, sliders, twisters, flappers, and jumpers in a comparatively short time. One of the more advanced walkers is detailed in the following illustration.

VBUG 1.5 "WALKMAN"
Single battery. 0.7Kg. metal/plastic construction. Unibody frame.
5 tactile, 2 visual sensors.
Control Core: 8 transistor Nv.
4 tran. Nu, 22 tran. motor.
Total: 32 transistors.
Behaviors:
• High speed walking convergence.
• Powerful enviro. adaptive abilities.
• Strong, accurate phototaxis.
• 3 gaits; stop, walk, dig.
• Backup/explore ability.

Figure 16.1. Walkman 1.5, an example of a motorized biomech made from cassette player components. (Source: *Living Machines*, 1994. Reprinted with permission.)

These machines "evolve" to take advantage of their own particular materials and control attributes in much the same way as nature has taken advantage of carbon and biological neural networks. As such, this robotic development technique is known as *biomorphic* research (from the Latin, meaning "of a living form") and the devices derived are *biomechs* (from a science fiction novel, the name of which is lost to memory). More than 200 robots of 30 different "species" have been built and studied, sometimes through three or more generations of improvements. One of these species involved nitinol-actuated machines, which were taken from initially primitive designs through four generations before other matters took precedence (research into nitinol machines was far from finished, but there is always something else that needs doing).

This chapter describes a few of the author's nitinol design attempts along biomorphic lines, and some minimal sure-fire circuits that seem to work well and for quite a while. This chapter does not contain a plan for a new form of Stiquito, but instead provides some ideas on how to make a variety of complete, independent, reliable nitinol mechanisms quickly, compactly, and at low cost. Nitinol is tough stuff to work with, but as with most new engineering fields, a little experience can yield top results.

The bottom line is that the only way to advance research in robotics is not with another computer simulation, but by building as many different robots as possible, advancing robot "genetics" through successive experimentation and display. This sounds like a good idea, but when you consider the amount of time and money normally involved in building a standard computer-based robot, even enthusiasts can get discouraged quickly. Nitinol walkers have the distinct advantage that a design can be tried, and if it doesn't work, little effort is lost but valuable experience is gained.

The bottlenecks are that controllers (usually microcomputers) and the power systems (usually off-board) make the robots, which generally weigh less than 10 grams, difficult to control, and they tend to overload the nitinol actuators, which then break. During this evolution, components are pushed well past their initial design specifications—but then progress is never achieved by following instructions. Other options must be possible, so the idea of biomech research is to allow exploration of neural and nervous control structures in 3-D space, with a hope of discovering the best types of adept robotic structures.

CONTROLLERS

Until a more effective alternative can be found, autonomous biomorphic machines must run under solar power. The problems are that conventional solar cells are typically too large, fragile, and inefficient to provide continuous energy to a motor system, and pure sunlight is not always available. A minimal biomech therefore needs a robotic "digestive system" that takes what little energy a solar cell does put out, integrates it, and delivers it to an actuator at reliable intervals. The cheapest and most effective circuit that can do this for nitinol (as of the time of this writing) is detailed in the following illustration.

Essentially, the Solarengine (covered under international patents) is a modified SCR design with supercritical feedback. In voltage-triggering mode (that is, with a 3-volt Zener

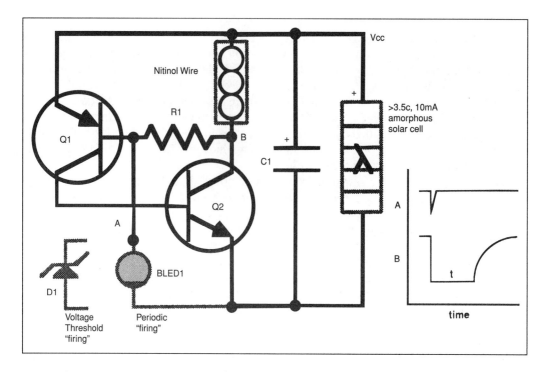

Figure 16.2. The nitinol Solarengine schematic (Nv–). (Source: *Living Machines*, 1994. Reprinted with permission.)

installed at point A), the Zener D1 starts clamping at Vcc –0.7v, eventually triggering Q1 as Vcc rises, triggering and latching Q2 into a low-impedance state so long as charge remains in capacitor C1. As the trigger event depends on perturbations in the transistor gains along the slow C1 charge path, the Solarengine can be considered an effective quasiperiodic oscillator; more so when considering variables in nitinol muscle load, inertia, and, of course, the variability of environmental light sources. Essentially, when the capacitor charges, the circuit fires, the nitinol contracts and expands, and the process repeats indefinitely.

Transistors

The transistors used are typically 3904 and 3906, but for higher power designs (that is, longer nitinol lengths), Q2 can be doubled up, or replaced with higher-current transistors like 2N2222 or equivalent, or high-gain NPN power transistors like the TIP31. The only transistors that seem to fail consistently are those used for surface-mount applications; they heat up too quickly under the currents required for nitinol actuators, which is a pity because SMT Solarengines would be small enough to mount 10 to a thimble.

Using FETs to drive nitinol is discouraged, as they do not work well at low voltages, and their high current delivery tends to fuse and strain nitinol excessively. They are expensive, too, but if you have the resources, do experiment. Good circuits are always in demand.

Resistor

The value of R1 determines how much current flows through the nitinol actuator. R1 should be between 1,000 and 100,000 ohms; it gives more nitinol contraction speed at lower values with less energy efficiency (that is, at low R1 values, the transistors will start to drain most of the energy, but the nitinol will contract faster). If 0.004-inch diameter nitinol is used, to get an equal contraction-versus-expansion coefficient, the following equation works pretty well.

R1 = 1000 x (nitinol length in centimeters)

For example, if your actuator uses 10 centimeters of exposed 0.004-inch diameter nitinol wire under tension, then an optimal R1 would be 10,000 ohms. Using this equation means the nitinol will contract and expand evenly and slowly. This extends the life of your nitinol muscle and means that the motions of all actuators in a multileg design can be calculated easily.

The value for C1 is dependent on the width and length of nitinol used, but it is typically greater than 30,000μF at 5 volts (using the microcapacitors used for memory backup in most computer electronics) and usually no more than 200,000μF per actuator (nominal 10-centimeter nitinol length).

Solar Cell

Amorphous solar cells are those found in most solar calculators, and are usually rated at 0.7 volts per individual cell segment. They deliver between 2.7 volts and 6 volts at 2 milliamps (room light) to 10 milliamps (full sunlight) and are sometimes cheaper and easier to obtain by destroying a calculator than they are to buy separately. As a rule, the higher the amorphous cell voltage, the more efficient the nitinol Solarengine; and the greater the current, the faster the triggering time.

The Solarengine circuit has such high off-state impedance that high solar cell current is not crucial; although with more current, the device will fire more frequently. The key is to have enough voltage for the circuit to operate. Nitinol Solarengine circuits have been made to trigger on as little as 1.2 volts, but the best range is between 3.5 volts and 6 volts; the nitinol contraction is directly proportional to the voltage in the main storage capacitor. Optimally, the best solar cells to use are those with five or more segments, and dimensions larger than 1.5 square inches. These are available commercially, but be sure to get cells with solderable leads.

Some work has been done with placing button-cell rechargeable lithium batteries across the solar cells. This works, but keep the cap in the circuit anyway, as lithium batteries don't handle high-current drains very well. With a bypass switch, your biomech can have either a slow-trigger independent mode, or a high-speed demonstration mode for those all-important times when there is no sun around.

Trigger Generators

To get the circuit to fire when a particular voltage is reached, a Zener diode trigger is recommended. The Zener should be rated at 1 volt below the solar cells' maximum voltage rating to get the best power efficiency, but can be any voltage (above 1.2 volt minimum) and still operate. To get higher trigger voltages, one or more 0.5-volt diodes can be put in series with the Zener (in forward bias), and the trigger voltage will increase appropriately.

The Zener can also be replaced by four or five regular diodes in series if necessary (forward biased), but more efficient solutions involve low-power circuits, such as digital watches (using the alarm outputs as a trigger), CMOS timers, voltage regulators, and blinking LEDs (BLEDs). The Zener circuit does drain current at a particular voltage level without triggering the transistors and thus needs a high-current (10 milliamp) solar cell to work effectively.

If a BLED is used as a trigger generator to fire the circuit (standard green BLEDs seem to work best, and are usually inexpensive), it replaces the Zener but will only trigger at 2.3 volts, and then only once a second. Again, to get higher trigger voltages, one or more 0.5-volt diodes can be put in series with the BLED (forward biased) and the trigger voltage will increase appropriately.

The higher the Solarengine's trigger voltage point, the more power is delivered into the nitinol actuator, but the fire frequency is slower. Fine-tuning is necessary to get optimal activation efficiency. Always be sure to breadboard your controller designs first.

NT SOLARENGINE CONSTRUCTION

The components of the nitinol Solarengine can be directly soldered together as detailed in the following illustration, and permutated as necessary based on your biomech's particular morphology. That is to say, you can fit one or more of these circuits into anything.

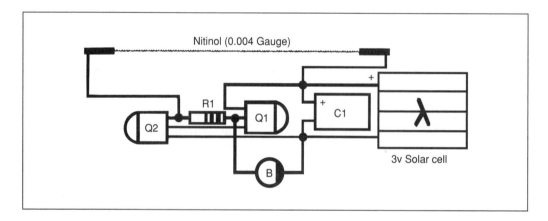

Figure 16.3. Nitinol Solarengine physical layout (example). Dots denote solder joins between components. (Source: *1995 BEAM Robot Games Rulebook.* Reprinted with permission.)

The advantages of the nitinol Solarengine design are small component count and adjustability, but mostly a very high off-state impedance until triggered. This means biomech designs can use very small solar cell arrangements that are robust against damage. This is compared to designs where massive, fragile cell arrangements are used just to supply the direct current necessary to contract the nitinol. Furthermore, Solarengine biomechs can be small and relatively independent of heavy power systems like tethers, heavy nicad batteries, power converters, or massive capacitor banks typical in robot designs. The Solarengine design gives the roboticist a lot of design latitude because all of the components can be mounted permanently on the machine itself.

Note that if you must tether your robot, take the cable from a dead computer mouse. Because it is soft, supple, lightweight, and long-lasting, it's pretty much ideal for the job, although it is still rather heavy and may result in nitinol breakage.

A TOUCH OF THEORY

With diode D1 removed and a coupling capacitor connected in its place, this circuit is, metaphorically speaking, like a biological motor neuron that amplifies the forward edge of input signal A into a pulse of variable length t resulting in mechanical energy and a signal output B. This signal output can be used to trigger other neurons either directly or through resistive connections (either inhibitory or excitatory). Considering that the output response results in a delay factor t that can be larger or smaller than the length of the input signal A, much like a biological neuron, the Solarengine circuit is also defined as the *Nv motor neuron (Nv)*. As there are two forms of this circuit (the one shown in the following figure and its electrical inverse, with the transistors and polarity swapped), there are two forms of Nv neuron: Nv+ and Nv–. Nv– is used more often simply because NPN transistors are electrically more capable than PNP for delivering efficient drive current. The common symbol for the Nv– motor neuron is shown in the figure, and is convenient when designing with groups of Solarengines for multileg designs.

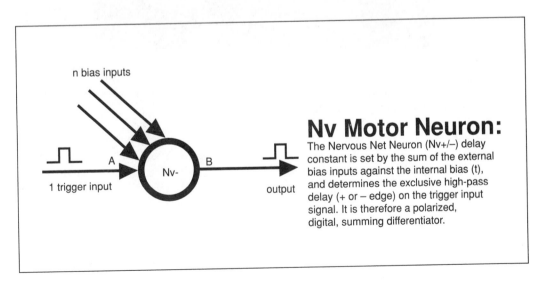

Nv Motor Neuron:
The Nervous Net Neuron (Nv+/–) delay constant is set by the sum of the external bias inputs against the internal bias (t), and determines the exclusive high-pass delay (+ or – edge) on the trigger input signal. It is therefore a polarized, digital, summing differentiator.

Figure 16.4. Symbol for the nervous net motor neuron (Nv–). (Source: *Living Machines*, 1994. Reprinted with permission.)

Interestingly enough, the biological motor neuron analogy is correct in several ways. Like biological motor neurons, the Nv Solarengine is self-contained, self-regulating, self-powered, and fires in an all-or-nothing manner only when the input trigger threshold is exceeded. Externally it looks like an oscillator, but its ability to keep performing under a wide variety of power, temperature, and load conditions gives it an almost biological robustness.

A further attribute is that the oscillation characteristics can vary wildly with minor parameter changes. Dynamic actuator loads or overpowered solar cells can make the Nv neuron fall into complex and interesting resonance patterns. This has advantages when different walking gaits and directions are desired for legged robots.

The two-transistor form of the Nv neuron is commonly called the BEAM Solarengine. It was developed in 1991 as the optimal starter circuit for first-time BEAM Robot Olympics competitors and works equally well with nitinol, small DC motors, speakers, and solenoids. The initial challenge of these Olympics is to build a self-contained device that covers a 1 meter distance in pure sunlight using only a 1.25-square-inch solar cell. It is the simplest implementation of a self-mobile nervous-net neuron, and is promoted because of its simplicity, ruggedness, and low cost.

Figure 16.5. An example of a solaroller design, Solaroller 1.2 (1991). Weight, 50 grams; length, 7.5 centimeters. Minimal components are necessary. (Source: *1994 BEAM Robot Games Rulebook*, LANL Press. Reprinted with permission.)

More complex forms of the Nv neuron involve digital circuits biased in unhealthy ways (that is, ways that usually let the smoke out). The properties of nervous-net clusters are still under study, but it has been seen that various Nv clusters have emergent properties that mimic many of the autonomic processes necessary for complex pseudobiological motions. Self-organizing and with broad behavior spectrums, Nv machines walk, adapt, and may soon see and even think. But the path of robot evolution starts with a single neuron, and in this case, the basic Nv Solarengine does the trick nicely.

The bottom line is that if you put nitinol Solarengines together in pairs, strings, or loops with capacitors, sensors, and/or resistors, you'll see some phase interactions that make excellent walking patterns. Such experimentation is left as an exercise for the reader.

Nitinol Actuators

The basic Stiquito leg is a brilliant design, but it does suffer from the disadvantage that the nitinol contraction down the leg causes the robot to waste a lot of effort pulling on itself rather than the outside world. This is what makes strong nitinol crimps necessary, and is suspected to be the main reason for nitinol snapping during operation. With a little remodeling, however, a new nitinol actuator can be devised with advantages over the standard leg. First, here are some basics for making high-quality crimps.

- Use clean brass or copper tubing for your crimp points instead of aluminum. These metals are stronger for holding onto the nitinol knots and have the advantage that you can solder to them.

- When preparing a crimp point for a nitinol wire, follow these steps carefully.

  1. Cut the tubing to the exact length: Put it on a paper surface, then roll a straight but dull knife blade over the cut point so the tube rolls back and forth under the blade until it either notches deeply around its diameter or snaps. A dull Exacto blade is recommended because, if a sharp blade is used, it's difficult to keep it in the same notch. This makes a clean cut around the tube which, when broken, will leave the end perfectly round.

  2. Ream out the inside edge of the crimp tube with a file, knife, or fine reaming tool. This removes any burrs in the metal that can cut your nitinol.

  3. Tie two knots in the nitinol around an upright pin so you can move the knots to the end of the wire length. When the knots are in position, remove the pin and pull the knots tight.

  4. Insert the nitinol in the crimp tubing and, using *flat-bladed* pliers (important!), crush the tubing flat over the nitinol knots. Repeat for both sides of the particular crimp to ensure the crimp is completely flat. (Caution: Do not use a hammer; you can split the tubing, cut the nitinol wire, and likely injure your thumb.)

You now have a crimp point that should be fairly strong. To mount the wire so it won't snap with wear, firmly hold the nitinol end with the same flat-bladed pliers and solder the end to the robot frame or printed circuit. It is vital that you maintain a firm hold; the pliers must absorb the excessive heat that would otherwise destroy your nitinol. Use heavy rubber bands around the plier handles to maintain pressure, and remove only after the tubing is cool enough to touch. Note that if your crimp tubing is very long (that is, the nitinol is more than 2 inches away from your solder point), then you don't have to be so careful. Still, it is better to be safe than sorry.

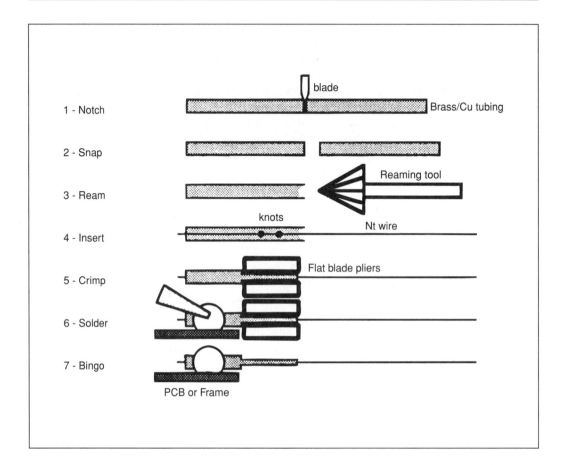

Figure 16.6. Seven steps to reliable nitinol crimps that you can solder.

To further ensure a reasonable nitinol lifetime, align the crimp tube on your robot frame so that the nitinol wire comes straight out of the crimp and not at an angle. This reduces stress on the wire, as nitinol does not work well around sharp corners (the inside of a nitinol bend contracts less than the outside of the bend, causing cracks to form in the wire).

LEG DESIGN

The elegance of the basic Stiquito leg is not just its simplicity but also its rigidity and compliance characteristics. Experiments using nitinol with handmade springs, levers, and pulleys have proved unsatisfactory because it is difficult to make a spring take any significant external load without gears, ratchets, or other complicated mechanics. Using spring-wire (or piano wire) for biomech frames allows for designs that can be formed into almost any shape, and fine-tuned as necessary.

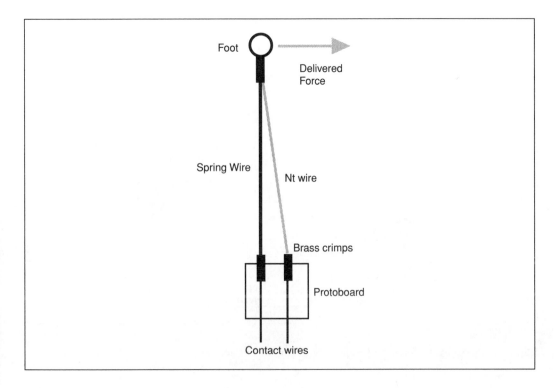

Figure 16.7. Standard Stiquito leg actuator.

The aforementioned crimp technique allows for reliable Stiquito leg designs, but there are still problems. Nitinol can be bought pretensioned, but its tensile characteristics will vary over its service life, and when you have to continually (and carefully) readjust the leg it can be a problem. And because the nitinol is out in the open, there is a strong likelihood that it will be hooked or damaged (especially where there is contact with other Nitinol creatures). Futhermore, as nitinol only has an 8 percent maximum contraction ability, the leg motions are often too small to make progress over rough surfaces.

The following design minimizes these problems. Called the Fishing Rod actuator (FRa), it is based on how fishing rods bend under a minor reel twist when the line is wound all the way in.

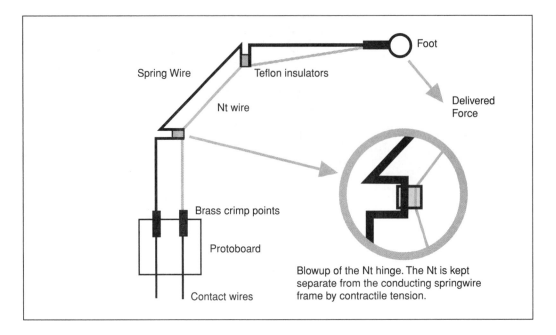

Figure 16.8. An example of the Fishing Rod (FRa) nitinol actuator.

The FRa leg is a single springwire frame bent into a U shape at regular intervals with hinge points where a small piece of Teflon insulation (normally stripped from 12-gauge wire, although 2-millimeter-diameter Teflon tubing can be bought separately) acts as a tendon sheath to keep the nitinol wire close and insulated from the conducting leg frame. The shape of the actuator frame must always be in an arc to keep the outward tension on the nitinol wire, and as such the wire is protected from whatever the biomech bumps into. A chief advantage of this design is that the leg can be twisted up and down from the initial plane of the bend and still keep its strength. That is, the main board contacts can remain flat, but the actuator can be bent forward or back from the curve plane to contact the surface as necessary.

Retensioning alignment is much easier with the FRa leg, as additional outward bending can be applied at any hinge while the other hinge(s) are bent in to ensure that the nitinol is not stressed (for this reason it is recommended that any FRa leg have at least two such hinge points equally spaced down its length). This lends to a further advantage that the foot's distance from the robot body can be adjusted in or out based on individual hinge angles. This is often important when adjusting the robot's legs to compensate for an uneven center of gravity.

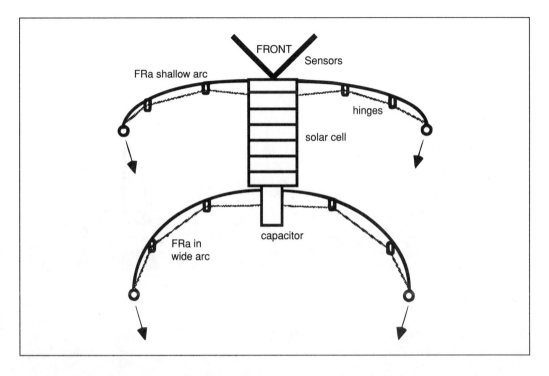

Figure 16.9. A sample nitinol "bug" robot design (top view) with FRa legs bent in different arcs. Such adjustment is often necessary to optimize the robot's walking ability. Built in 1994 as the third of the Solar Nitewalker series of biomech robots, its legs are first made flat, then twisted down to contact the feet with the floor.

The depth of the hinge brackets is important. Too deep and the leg becomes flimsy and the nitinol does not contract sufficiently to create a good motion. Shallow, squarish hinges are best and the rounder the leg shape, the better the foot travel.

Be sure to use real Teflon tubing at the hinge points. This is important for three reasons: (1) the nitinol tendon can slip past the Teflon easily; (2) Teflon is capable of taking much higher temperatures than the nitinol can deliver without melting; and (3) Teflon naturally forms to the nitinol wire, reducing the stress normally associated with running nitinol around sharp corners.

The closer the nitinol can travel around the FRa leg, the better the contractile distance. By a simple calculation of a difference of squares, it can be seen that if the nitinol were to follow around a perfectly circular leg, a small nitinol contraction would result in significant increase in effector travel while still retaining the necessary compliance to keep the nitinol safe. The problem is that the effector is now wrapped back around to its source, but even this can be made useful. The following snail-like biomech, called *Nitecrawler 1.0*, uses this technique to optimal advantage.

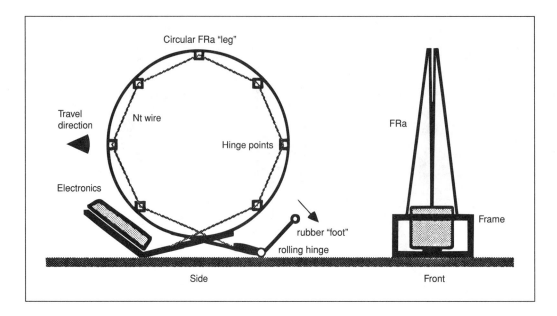

Figure 16.10. A design schematic of Nitecrawler 1.0.

The springwire frame is a long V-shape folded back onto itself with hinge points at 45-degree increments. A single Nv Solarengine in the head contracts 15 centimeters of 0.004-inch diameter nitinol, pushing the trailing foot backward until the rubber foot moves down on its hinge, hitting the walking surface and moving the entire unit forward. When the frame expands, the hinge lifts the foot and the device slides forward into its new position; the forward weight of the electronics and solar cell helps the process.

Solar powered and fragile, the Nitecrawler was nonetheless a great mover, traveling more than 3 centimeters with every "step". The motion derived was much farther than one would expect from 15 centimeters of nitinol, and during construction and operation, many other designs were envisioned: inchworms, rotary floppers, snap-action hoppers, starfish, swimmers, and so on.

The FRa leg is, of course, not the optimal nitinol actuator design, but it is an interesting one that is simple, effective, and fairly robust. It has been proven by experiment that when coupled with the brass-crimp technique, the FRa leg can allow the building of machines with significant expected lifetimes, perhaps years.

Construction

For best results, keep the machine design simple enough so that large printed circuit boards are not necessary. Nitinol's main advantage is its excellent power-to-size ratio. Avoiding the additional weight of printed circuit boards (or tethers) means less strain on the fragile nitinol actuators. Where possible, also avoid the use of wire-wrap sockets or big chip holders; these are expensive, complicate your design, and will eventually fail from oxidization. Test out your electronic design on a breadboard first (for several days if possible) and then solder the components in. Assume they are there to stay; if you want to try again, build an entirely new robot rather than cannibalize previous ones for parts.

In the robot evolutionary process, it's important to keep your old successes (or mistakes) around for comparison and demonstration. This is always useful when someone says, "Have you tried —?" and it also keeps you from the curse of reinvention.

What is important for Nitinol designs is a knowledge of materials. It's important to know what your machine will be made of and *how* for two major reasons: (1) if your machine is to last, it must be made tough, and (2) because things get dropped, pounded, and stressed, it must be made flexible. These are not incompatible requirements; there are ways to design nitinol machines that are strong and resilient. The larger your device, the more resilient it must be to survive.

Remember that solder is an excellent structural adhesive for small robot designs. It is adjustable, clean, and conductive, and won't weld fingers together like some glues. Water-soluble flux solder is best; a quick trip through a dishwasher will give any nitinol robot a fine shine.

THE FUTURE OF NITINOL

Nitinol robot evolution could be the best start on a long road toward a future filled with independent, useful machines. Nitinol is great, but it has one major problem: It consumes a lot of power, especially when you realize how much sustained current is necessary to heat up the wire and get it to contract. Nominally, nitinol has only about 2 percent efficiency, which would be fine if it had a flat power profile. But, like an inordinately heavy motor, the starting current is excessive: too high for too long, and suddenly your battery is nowhere near as adequate as you had first imagined.

The second problem with nitinol is that it strains. Too much current too fast, and the metal becomes a short-lived fuse. Too much load too fast, and the same thing happens. This is fine for low-load applications like moving cardboard pendulums, but when used for legs or manipulators against a real world, you must to take into account the intractable nonlinear loads that, say, a rocky terrain will inflict on your robot.

These are the current goals of most legged robots: not just to do as well as wheels or tank treads, but better. A good standard is a machine that can walk over any obstacle its own height; up complex inclines in excess of 45 degrees; and, ideally, right up a vertical wall. A further goal for legs is not just to be movers, but also to be actuators so that when the robot gets where it's going, it has the tools necessary to do a job.

There are other reasons why legs are so much better than wheels for coping with a complex world over a long period of time. Losing a leg isn't as serious as losing a wheel. Wearing off an inch of tread on a wheel is disastrous, but wearing a few inches off a long set of legs is not a problem. There is also evidence that legs allow machines to communicate at a low level through structural interactions that form coupling (two robots walking side by side with their legs meshed will tend to walk in sync, for example). The list goes on but research is only just beginning.

So what about the long-term future of nitinol walkers? Why do we need them? Well, besides being exciting and giving some insight into the nature of biological interaction, they could become a necessity. Today most cleanup, pesticide, and construction operations are handled with chemicals that are hard on the environment. But if the same operations could be done with solar-powered, disposable, biodegradable robots that won't contaminate the groundwater—robots that stop when we tell them to, but keep working as long as the sun shines—there may be a future for nitinol walkers indeed.

Grass-cutting, tree-pruning, seed-planting, bug-killing, garbage-sorting, window-cleaning, pipe-unclogging, floor-scrubbing, dust-collecting, radiation-harvesting, mine-clearing, space-exploring machines could become as commonplace as the light bulb. Designed for particular application environments, there will be no fear of them "taking over" because they'll only be smart enough for the job they are optimized to fill. Because they cannot reproduce, they'll never get out of control. Because we can make them small and weaker than ourselves, we'll never have cause to fear them. And if we can build them cute enough, by gosh, they'll sell.

To work, these robots will have to come in thousands of shapes and sizes, made from materials and robobiological controllers that optimize their power, strength, function, and compliance attributes. They will live on light, which is cheap and readily available, so nothing in nature need compete with them. They will live parallel to both plants and animals, doing tasks for both but obstructing neither. And when they finally break down, the silicon, metal, and trace elements of which they are composed will return to the earth, or be recycled into something new.

To make the small ones, the flat ones, the flying ones, or the ones that can go where motors cannot, nitinol biomech designers would be in demand. They would be a necessary part of the new field of parallel-life robotics, a systematic approach to building independent machines capable of doing long-term work with incomplete or inexact work specifications under environmental duress.

Despite current limitations, through experimentation and experience Nitinol robotics can realize the promise of all these things. Some assembly required.

REFERENCES

1. Degaris, Hugo. 1993. *Genetic programming: GenNets, artificial nervous systems, artificial embryos.* Wiley Publishers.

2. Hasslacher, B., and M.W. Tilden. 1995. Living machines. In *Robotics and autonomous systems: The biology and technology of intelligent autonomous agents*, L. Steels, ed. Elsivier Publishers.

3. Rietman, Ed. 1994. *Genesis redux: Experiments creating artificial life.* Windcrest/ McGraw Hill, pp. 295-301.

4. Tilden, M.W. 1995. *BEAM 4: The international BEAM robot olympic games rulebook.* Los Alamos, N.M.: Los Alamos National Laboratory Press.

5. Tilden, M.W. 1995. Biomorphic robots as a persistent means for removing explosive mines. *Proc. Symp. Autonomous Vehicles in Mine Countermeasures.*

Technical contents of this chapter are covered under existing and pending international patents.

Appendix A

Author Biographies

James M. Conrad received his bachelor's degree in computer science from the University of Illinois, Urbana, and his master's and doctorate degrees in computer engineering from North Carolina State University. He is currently an engineer at Ericsson, Inc., and an adjunct professor at North Carolina State University. He has served as an assistant professor at the University of Arkansas and as an instructor at North Carolina State University. He has also worked at IBM in Research Triangle Park, North Carolina, and Houston, Texas; at Seer Technologies in Cary, North Carolina; at MCI in Research Triangle Park, North Carolina; and at BPM Technology in Greenville, South Carolina. Conrad is a member of the Association for Computing Machinery, Eta Kappa Nu, and IEEE Computer Society, and is a senior member of IEEE. He is the author of numerous journal articles and conference papers and has written two books on robotics, parallel processing, artificial intelligence, and engineering education. Conrad can be reached at the following address:

Ericsson, Inc.
7001 Development Drive
Research Triangle Park, NC 27709
Phone 919-472-6178
Fax 919-472-6515
E-mail jconrad@acm.org (preferred contact method)

Jonathan W. Mills received his doctorate in 1988 from Arizona State University. He is currently an associate professor in the Computer Science Department at Indiana University and director of Indiana University's Analog VLSI and Robotics Laboratory, which he founded in 1992. Mills invented Stiquito in 1992 as a simple and inexpensive walking robot to use in multi-robot colonies, and with which to study analog VLSI implementations of biological systems. In 1994 he developed the larger Stiquito II robot, which is used in an eight-robot colony in his laboratory. Since 1992 Indiana University has distributed more than 3,000 Stiquito robots, leading to the idea for this book.

Mills is currently researching biological computation in the brain using tissue-level models of neural structures implemented with analog VLSI field computers. Field computers offer a powerful but simple paradigm for adaptive robotic control. They are small and light enough to be carried by Stiquito, yet still perform sensor fusion and behavioral control.

Mills has written a series of papers on his analog VLSI and robot designs, and has one patent with several others pending and applied for on his work. He also freely admits that Stiquito is just the start of what he hopes will be a series of improved and functional miniature robots, and encourages the readers of this book to be inspired to design and build them. Mills can be reached at the following address:

215 Lindley Hall, Computer Science Department
Indiana University
Bloomington, IN 47405
Phone 812-855-6486; Fax 812-855-4829
E-mail stiquito@cs.indiana.edu

Christopher A. Baumgartner received his bachelor's degree, magna cum laude, in computer science and engineering from the University of Toledo in 1993. He is a member of the Association for Computing Machinery.

Joyce Binam graduated from the University of Arkansas in 1994 with a bachelor's degree in computer systems engineering. While attending college, she received the GE Foundation Research Award. Binam works in northwestern Arkansas, where she is a system analyst in the Information Systems Department of St. Mary's Hospital.

John K. Estell received his bachelor's degree, summa cum laude, in computer science and engineering from the University of Toledo in 1984. Awarded both a National Science Foundation Graduate Fellowship and a Tau Beta Pi Fellowship, he earned his master's and doctorate degrees in computer science from the University of Illinois at Urbana-Champaign. From 1991 to 1996 Estell was an assistant professor in the Electrical Engineering and Computer Science Department at the University of Toledo, and he is currently an associate professor of computer science at Bluffton College.

Estell's research interests include microcomputer systems, interface design, and Web-based program development tools. His work with nitinol-propelled robots has led to several papers and presentations, one of which received the 1994 John A. Curtis Lecture Award from the American Society for Engineering Education. Estell is a member of ASEE, the Association for Computing Machinery, Eta Kappa Nu, IEEE, IEEE Computer Society, Phi Kappa Phi, and Tau Beta Pi. When not working with computers or robots, he can often be found leading hikes for the Sierra Club or performing on one of his reed organs. Estell can be reached at the following address:

Computer Science, Box 695
Bluffton College
Bluffton, OH 45817-1196
Phone 419-358-3375; Fax 419-358-3323
E-mail estell@bluffton.edu

Greg Evans graduated from the University of Arkansas with a degree in computer systems engineering. He works for the Monsanto Company in St. Louis, Missouri, as a client server technologist. His interests include aquariums and cheering on the Razorbacks in any sport. Evans can be reached at the following address:

Monsanto
800 North Lindbergh
St. Louis, MO 63167
Phone 314-694-6562
E-mail: glevan@ccmail.monsanto.com

Roger G. Gilbertson received his bachelor's degree from Hampshire College in 1984, where he combined his wide-ranging interests to produce film and computer studies of the walking patterns of legged animals. This research led to designs for robot mechanisms that adopted many of the structures of living creatures, but these designs proved difficult to build without muscle-like actuators. In 1986 he encountered samples of a Japanese-made shape-memory alloy wire, and, knowing immediately how he would use it, began building his mechanisms, eventually working with the manufacturer.

In 1991 Gilbertson wrote *The Muscle Wires Project Book*, now in its third edition, which has introduced tens of thousands of individuals to the practical uses of shape-memory alloy wires. He has great interest in the evolution and synthesis of technologies for creating a supportive and sustainable future. Gilbertson resides in Marin County, California. He can be reached at the following address:

Mondo-tronics, Inc.
4286 Redwood Hwy. #226
San Raphael, CA 94903
Phone 415-491-4600; Fax 415-491-4696
E-mail info@mondo.com

Lance Hankins is a student at the University of Arkansas, working on his master's degree in computer engineering. He has worked on predicting the stock market with neural networks and porting the RAPs Interpreter from Lisp into the object-oriented paradigm using C++. He has been funded as a teaching assistant conducting labs in digital techniques, and as a research assistant working on intelligent architectures for robotics. He spent a summer at the Jet Propulsion Laboratory (Pasadena, California) working in the Software Quality Assurance Group. His thesis involves researching the cognitive issues of object-oriented programming and building an integrated development environment that will address several of the issues arising from this research. Hankins can be reached at the following address:

University of Arkansas
Computer Systems Engineering
313 Engineering Hall
Fayetteville, AR 72701-1201
Phone 501-575-6036
E-mail lwh@engr.uark.edu

Quan D. Luong received his bachelor's degree, cum laude, in computer science and engineering from the University of Toledo in 1993. Luong is a member of the Golden Key National Honor Society.

Paulo W.C. Maciel was born in Rio de Janeiro, Brazil, in 1958. He received a bachelor's degree in electronics and power systems in 1981 and a master's degree in computer science in 1986 from the Catholic University of Rio de Janeiro. After earning his bachelor's degree he joined the Brazilian National Telecommunications Company as a trainee in a data networks training program, and later became a software engineer. In 1995 he earned his doctorate in computer science, with a focus on computer graphics, from Indiana University in Bloomington. He is currently a research-and-development graphics software engineer at Hewlett-Packard's software lab in Corvallis, Oregon. Maciel can be reached at:

Hewlett-Packard, Workstations Group Corvallis
1000 NE Circle Blvd., Mailstop 524A
Corvallis, OR 97330-4239
Phone 503-715-2069
E-mail pmaciel@cv.hp.com

Susan A. Mengel has worked in the area of artificial intelligence for many years. She has used neural networks to represent the student model in intelligent tutoring systems; to predict the stock market; and for preliminary target recognition work. She spent a summer helping a consulting firm become knowledgeable in artificial intelligence techniques. While at the University of Arkansas, she was funded as principal investigator in the area of cooperative mobile robotics under a NASA Arkansas space grant. She was a NASA Summer Faculty Fellow at the Jet Propulsion Laboratory (Pasadena, California) in the Software Quality Assurance Group looking at validation techniques for autonomous and nonautonomous spacecraft systems. At Texas Tech University, she is actively engaged in research for multimedia instruction and software engineering. She is a member of IEEE, ACM, International Neural Network Society, Society of Women Engineers, and ASEE. She is active in engineering education and is the newsletter editor for ASEE's Educational Research and Methods Division (http://aln.coe.ttu.edu/erm/). Mengel can be reached at the following address:

Texas Tech University
Computer Science
Box 43104
Lubbock, TX 79409-3104
Phone 806-742-3527; Fax 806-742-3519
E-mail mengel@ttu.edu

Roger Moore is a student at the University of Arkansas, working on his master's degree in computer engineering. He has worked on predicting the stock market with neural networks and using fuzzy logic to help robots cooperate on tasks. He has been funded as a teaching assistant helping teach a data structures course and as a research assistant working on intelligent architectures for robotics. He spent a summer at the Jet Propulsion Laboratory (Pasadena, California) working in the Software Quality Assurance Group. Moore can be reached at the following address:

3560 Alma Rd., #2123
Richardson, TX 75080
E-mail moore@gar.esys.com

Timothy A. Muszynski received his bachelor's degree, magna cum laude, in computer science and engineering with a minor in mathematics from the University of Toledo in 1993. He is currently working as a technical aide at the Davis Besse Nuclear Power Plant in Oak Harbor, Ohio. His interests include operating system design, multiprocessing and multiuser environments, and cross-platform system applications. Muszynski is a member of Eta Kappa Nu, Pi Mu Epsilon, and the Golden Key National Honor Society.

Mohan Nanjundan received his bachelor's degree in electrical and electronics engineering from Bharathiar University, Coimbatore, India, in 1988, and his master's degree in electrical engineering from the University of Arkansas at Fayetteville in 1993. He has worked as a design engineer at Jaisun and Hutchison Private Controls, Ltd. in India. He is currently working as an applications engineer at Hewlett-Packard in California. Nanjundan can be reached at the following address:

Hewlett-Packard
1601 California Ave.
Palo Alto, CA 94304
Phone 415-857-4567
E-mail mohan_nanjundan@sid.hp.com

Thomas A. Owen received his bachelor's degree in computer science and engineering from the University of Toledo in 1994. His interests include hardware design, interface design, and embedded systems applications.

Shyam Pullela received his bachelor's degree in computer science from the Regional Engineering College in Warangal, India. He received master's degrees from the Indian Institute of Technology in Kanpur, India, and from Indiana University. He is currently working as a hardware design verification engineer at the Convex Technology Center of Hewlett-Packard in Richardson, Texas. He is also a doctoral candidate at Indiana University. Pullela can be reached at the following address:

7650 McCallum Blvd., #1104
Dallas, TX 75252
Phone 972-267-0890
E-mail pullela@convex.com

Matthew C. Scott received his bachelor's degree in computer science from Indiana University and is now pursuing a master's degree in electrical engineering at the University of Arizona. He is currently a senior design automation engineer and Webmaster for Burr-Brown in Tucson, Arizona. His obsessions include genetic algorithms, neural networks, and the philosophy of consciousness. Scott may be reached at:

http://www.ece.arizona.edu/~scottm
3528 E. 2nd St., Apt. 87
Tucson, AZ 85716
Phone 520-746-7467
E-mail scott_matthew@bbrown.com

Steven R. Snodgrass received his bachelor's degree in computer science and engineering from the University of Toledo in 1993. His interests include parallel architectures, operating systems implementation, and object-oriented programming. Snodgrass is a member of Sigma Pi Sigma.

Craig A. Szczublewski received his bachelor's degree in computer science and engineering from the University of Toledo in 1994. His interests include embedded systems design and real-time programming applications.

Jason A. Thomas received his bachelor's degree in computer science and engineering from the University of Toledo in 1993. He is currently working as a software engineer developing computer vision inspection systems.

Mark Tilden was born in England and educated at the University of Waterloo in Canada in systems engineering, subsequently working in the Math Faculty of that university for seven years as a systems engineer. He is currently a research scientist at Los Alamos National Laboratory in New Mexico, furthering the study and application of biomech robots for home, education, and industry. Tilden has written several papers on the applications and theory behind his Nervous Network technology and a book on the International BEAM Robot Games, which he started in 1991. Tilden can be reached at the following address:

MSD454, LANL
Los Alamos, NM 87545
E-mail mwtilden@math.uwaterloo.ca

Appendix B

An Analog Driver Circuit for Nitinol-Propelled Walking Robots

Several of the chapters in this book discuss advanced controllers for nitinol-propelled walking robots. The digital controllers described, however, are not necessary to make the Stiquito robot described in Chapter 2 walk. This appendix contains a schematic and brief instructions on how to build an analog controller. The controller is tailored for the Stiquito robot, but can be extended to other robots that use nitinol.

The schematic shown in Figure B.1 generates enough current to contract up to three nitinol actuators. To control all six Stiquito legs, you will need two of these circuits. Each circuit operates independently of the other, and controls the three actuators in a tripod gait (refer to the manual controller circuit in Chapter 2 for details on the tripod gait).

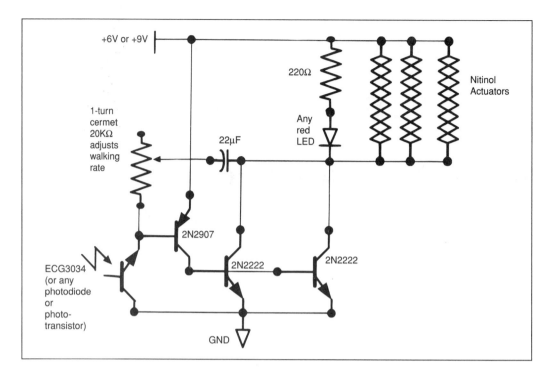

Figure B.1. Three-actuator nitinol driver circuit (©Jonathan W. Mills, 1992, 1994, 1996).

The controller circuit consists of three main parts.

1. The actuator components
2. The current-driving components
3. The switching/sensing components

The circuit will work with a 6-volt or a 9-volt battery. You can find the parts at any electronics supply store. Refer to the supplier list in Appendix C for nationally listed suppliers.

The parts needed to build the analog controller (with two of the circuits shown in Figure B.1) are as follows:

1	Battery connector
2	220Ω resistors
2	Red LEDs
4	2N2222 NPN transistors
2	2N2907 PNP transistors
2	1-turn 20KΩ cermet
2	ECG 3034 photodiode (or any photodiode or photo resistor)
2	22μF capacitors
1	Perfboard (plated holes are recommended)
	Miscellaneous soldering/wiring supplies

The photodiode provides a variable amount of resistance to allow the nitinol actuators to fire slowly or quickly. Mount the photodiodes or phototransistors on the end of long wires ("antennae") to get the best sensitivity to light for good turning and following response (see Figure B.2).

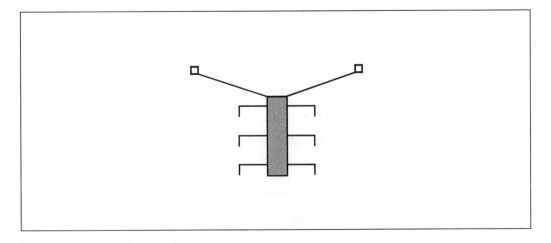

Figure B.2. Example of Stiquito with photodiode antennae.

Remember to adhere to common safety practices when building this circuit. Wear safety glasses and use extreme caution when working with a hot soldering iron.

The circuit in Figure B.1 will exhibit a light-following behavior. To make Stiquito steer away from light, simply wire the photodiodes to the other side of the body. That is, the diode on the antenna should be placed on the side of Stiquito that has two legs actuated in the tripod gait.

Instead of building a controller that follows light, you might want a controller that has Stiquito walk in a straight line (or as close to a straight line as possible). To do this, simply replace the photodiode with a resistor that has a large value (like 100KΩ). The cermet can also be replaced with a 6KΩ resistor. This new circuit is shown in Figure B.3.

The parts needed to build the analog controller (with two of the circuits shown in Figure B.3) are as follows:

1	Battery connector
2	220Ω resistor
2	Red LEDs
4	2N2222 NPN transistors
2	2N2907 PNP transistors
2	6KΩ resistors (or similar valve, dependent on your particular Stiquito)
2	100KΩ resistors
2	22μF capacitors
1	Perfboard (plated holes are recommended)
	Miscellaneous soldering/wiring supplies

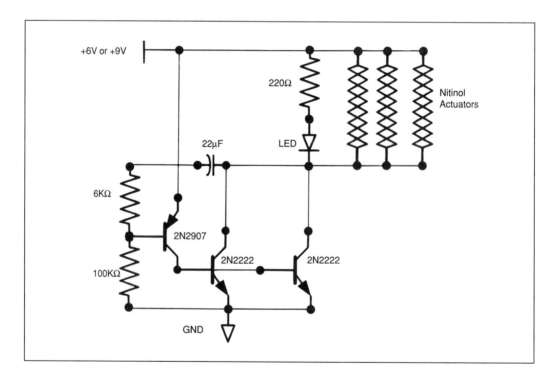

Figure B.3. Alternate analog controller for a robot that walks in a straight line.

One final note is on the use of this driver circuit for walking robots with two ranges of motion (like Stiquito II). The vertical drivers should not be connected unless additional circuitry is added to disable the horizontal drive circuitry. This is necessary because the oscillator drivers are not synchronized with each other. Without any circuitry to link the forward-backward motion of a leg with the raising and lowering of the same leg, you cannot be assured that the leg will be touching the ground when it is moved forward.

Appendix C

Sources of Materials for Stiquito

This appendix lists some suppliers of Stiquito parts, robotics kits, and electronics. You may find many Stiquito supplies and tools at your local hobby supply store. Check local electronics supply stores for other supplies. This section is not an endorsement of these companies, but is provided to make your supplier search easier.

All Electronics Corporation
P.O. Box 567
Van Nuys, CA 91408
Phone 800-826-5432, 818-904-0524; Fax 818-781-2653
E-mail allcorp@allcorp.com; Web address http://www.allcorp.com
Surplus dealer of boards, components, and assemblies.

Artificial Creatures
22 McGrath Hwy., Ste. 6
Somerville, MA 02143
Phone 617-629-0055; Fax 617-629-0126
E-mail art@isr.com; Web address http://www.isr.com
A subsidiary of IS Robotics. Supplies small mobile robots for research and education.

Digi-Key
701 Brooks Ave. South
P.O. Box 677
Thief River Falls, MN 56701-0677
Phone 800-344-4539 or 218-681-6674; Fax 218-681-3380
Web address http://www.digikey.com

Dynalloy, Inc.
3194-A Airport Loop Drive
Costa Mesa, CA 92626-3405
Phone 714-436-1206; Fax 714-436-0511
E-mail flexinol@dynalloy.com; Web address http://www.dynalloy.com
Supplier of nitinol wire, trade named Flexinol. Flexinol can be ordered in lengths of
1 meter or more.

Edmund Scientific
101 E. Gloucester Pike
Barrington, NJ 08007-1380
Phone 609-573-6250; Fax 609-573-6295;
International phone 609-573-6263; Fax 609-573-6882
Web address http://www.edsci.com
Sells optical components, science kits, surplus motors, and robot kits.

Hamilton Hallmark
10950 Washington Blvd.
Culver City, CA 90232
Phone 800-332-8638, 310-558-2494; Fax 800-257-0568
Distributor for many semiconductor manufacturers.

IEEE Computer Society Press
10662 Los Vaqueros Circle
P.O. Box 3014
Los Alamitos, CA 90720-1314
Phone 800-CS-BOOKS (800-272-6657) or 714-821-8380; Fax 714-821-4641;
E-mail cs.books@computer.org; Web address http://www.computer.org/cspress
Source for Stiquito books and kits.

IS Robotics
22 McGrath Hwy., Ste. 6
Somerville, MA 02143
Phone 617-629-0055; Fax 617-629-0126
E-mail art@isr.com; Web address http://www.isr.com
Source for research robots and sensor systems.

Jameco Electronic Components
1355 Shoreway Rd.
Belmont, CA 94002-4100
Phone 800-831-4242 or 415-592-8097; Fax 415-592-2503
E-mail sales@jameco.com or international@jameco.com
Web address http://www.jameco.com
Supplier of electronic components.

K&S Engineering
6917 W. 59th St.
Chicago, IL 60638
Phone 773-586-8503; Fax 773-586-8556
Supplier of music wire and aluminum, copper, and brass tubing. Minimum order of
$80.00. (Hobby shops also carry these items.)

LEGO DACTA
Lego Educational Department
P.O. Box 1600
Enfield, CT 06083
Phone for education purchases: 800-527-8339
Phone for retail purchases: 800-243-4870 or 860-749-2291
Web address http://www.lego.com/learn
Sells components needed for quickly building robot prototypes; educational department sells primarily to schools.

Micro Fasteners
110 Hillcrest Rd.
Flemington, NJ 08822
Phone 800-892-6917 or 908-806-4050; Fax 908-788-2607
E-mail microf@blast.com; Web address http://www.microfasteners.com
Supplier of brass #0 screws, nuts, and washers used on the Stiquito body.

Mondo-tronics
4286 Redwood Hwy., #226
San Rafael, CA 94903
Phone 800-374-5764 or 415-491-4600; Fax 415-491-4696
E-mail info@mondo.com; Web address http://www.robotstore.com
Supplier of nitinol and Flexinol wire, robots, robotic books, videotapes, even robotic artwork.

Mouser Electronics
2401 Hwy. 287 N.
Mansfield, TX 76063-4827
Phone 800-34-MOUSER (800-346-6873) or 817-483-5712; Fax 817-483-0931
E-mail sales@mouser.com; Web address http://www.mouser.com
Wide selection of electronic components. Regional distribution centers; will fax detailed specifications. Accepts small orders.

New Micros, Inc.
1601 Chalk Hill Rd.
Dallas, TX 75212
Phone 214-339-2204; Fax 214-339-1585
Single-board computer uses MC68HC11 chip; Forth language in ROM.

Newark Electronics
US and Canada catalog requests: 800-4-NEWARK (800-463-9275)
Central and South America 915-772-3192
Pacific Rim 619-691-0141
Europe, Middle East, Africa +44 (1420)543304
Check telephone directory for a local sales office.
Distributor of electronic components.

Parts Express
340 E. First St.
Dayton, OH 45402-1257
Phone 800-338-0531 or 937-222-0173; Fax 937-222-4644
E-mail xpress@parts-express.com; Web address http://www.parts-express.com
Supplier of electronics, tools, hardware, and supplies.

Plastruct, Inc.
1020 S. Wallace Pl.
City of Industry, CA 91748
Phone 818-912-7017 or 800-666-7015; Fax 818-965-2036
Supplier of plastic stock used in Stiquito II, Boris, and SCORPIO. Plastruct offers a
30 percent educational discount to instructors and schools using a purchase order.

Radio Shack
Phone 800-THE-SHACK
Web address http://www.tandy.com
National chain; consult telephone directory for nearest dealer. Offers a variety of
electronic components from local distributors. For mail order, see Tandy Electronics.

Micro-Robotic Supply, Inc.
101 Pendren Place
Cary, NC 27513-2225
E-mail sales@stiquito.com
Web address http://www.stiquito.com
Supplier of brass screw kits, Stiquito repair kits, and Stiquito accessories.

Small Parts, Inc.
13980 NW 58th Court
P.O. Box 4650
Miami Lakes, FL 33014-0650
Phone 800-220-4242 or 305-558-1255; Fax 800-423-9009 or 305-558-0509
Supplier of metal, plastics, tools, and hardware.

Solarbotics
179 Harvest Glen Way NE
Calgary, AB
Canada T3K 3J4
Phone 403-226-3783; Fax/alternate 403-226-3741
A source for solar-powered robot kits.

Tandy Electronics
National Parts Division
900 E. Northside Dr.
Fort Worth, TX 76102
Phone 800-322-3690; Fax 817-332-4216
Web address http://www.tandy.com
Parent company of Radio Shack; distributes (by mail order) a wider variety of parts than
are available in Radio Shack stores.

Appendix D

Technical Characteristics of Flexinol Actuator Wires

Flexinol Actuator Wires are small-diameter wires that contract like muscles when electrically driven. Smaller than motors or solenoids, cheaper and generally easier to use, these wires perform physical movement across an extremely wide variety of applications.

Table of Contents

Dynalloy, Inc.
Makers of Dynamic Alloys
3194-A Airport Loop Drive
Costa Mesa, CA 92626-3405
Phone 714-436-1206
Fax 714-436-0511
E-mail flexinol@dynalloy.com
Web address http://www.dynalloy.com

NICKEL-TITANIUM ALLOY PHYSICAL PROPERTIES

1. Density

   0.235 lb/in³
   6.45 gr/cm³

2. Specific Heat

   0.20 BTU/lb°F
   6-8 cal(mol.°C)

3. Melting Point

   2282 °F
   1250 °C

4. Heat of Transformation

   10.4 BTU/lb

5. Thermal Conductivity

   10.4 BTU/hr-ft-°F
   0.05 cal(cm-°C-sec)

6. Thermal Expansion Coefficient
   Martensite

   3.67 x 10⁻⁶/°F
   6.6 x 10⁻⁶/°C

   Austenite

   6.11 x 10⁻⁶/°F
   11.0 x 10⁻⁶/°C

7. Electrical Resistivity
   Martensite   421 ohms/cir mil ft (approx.)
   Austenite   511 ohms/cir mil ft (approx.)

8. Linear Resistance (approx.)
   .003 inch diameter wire   4.3 ohms/inch
   .005 inch diameter wire   1.7 ohms/inch
   .006 inch diameter wire   1.25 ohms/inch
   .010 inch diameter wire   0.44 ohms/inch

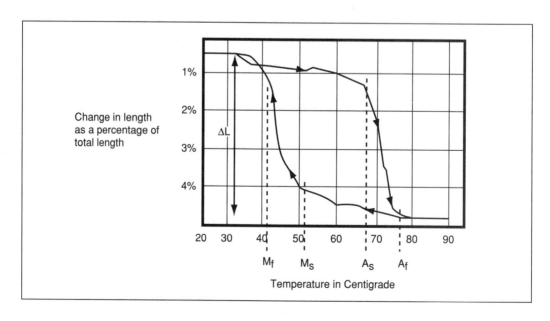

Figure D.1. Change of length of Flexinol with change in temperature while constant tensile stress is applied to the wire.

INTRODUCTION

Flexinol is a trade name for shape-memory alloy actuator wires. Made of nickel-titanium, these small-diameter wires contract like muscles when electrically driven. This ability to flex or shorten is characteristic of certain alloys, which dynamically change their internal structure at certain temperatures. The idea of reaching higher temperatures electrically came with the lightbulb, but instead of producing light, these alloys contract by several percent of their length when heated and can then be easily stretched out again as they cool to room temperature. As with a lightbulb, both heating and cooling can occur quite quickly. The contraction of Flexinol actuator wires when heated is opposite to ordinary thermal expansion, is larger by a hundredfold, and exerts tremendous force for its small size. The underlying technology that causes the effect is discussed in Section 5. The main point is that movement occurs through an internal "solid state" restructuring in the material that is silent, smooth, and powerful.

This effect can be used in many ways. The viable applications are too numerous for any single listing. A safe assumption is that any task requiring physical movement in a small space with low to moderate cycling speeds is something that most likely will be better done with actuator wires. Many of the tasks currently being done with small motors or solenoids can be done better and cheaper with Flexinol actuator wires. Since the actuator wires are much smaller for the work they do, a number of new products and improved designs on existing products are readily accomplished.

For new users of Flexinol actuator wires Dynalloy, Inc. strongly recommends that an overview of what can be done first be established. This can be done by obtaining one of the Dynalloy, Inc. kits made for such familiarization. Second, new users should consider obtaining from Dynalloy, Inc. or other consultants a "proof of concept" working model. This is not only useful as an internal marketing and sales tool, it also helps the new user to see how those with more experience approach the specific task at hand. Knowing this provides immeasurable insight into how to proceed and helps reduce the redundancy of reinventing existing technics. One can always improve on existing methods, and sufficient legal and other safeguards can be readily employed to ensure protection of proprietary ideas.

SECTION 1. MOVEMENT

The movement or stroke of Flexinol actuator wire is measured as a percentage of the length of the wire being used and is determined, in part, by the level of stress one uses to reset the wire, or to stretch it in its low-temperature phase. This opposing force used to stretch the wire is call the *bias force*. In most applications, the bias force is exerted on the wire constantly, and on each cycle as the wire cools, it is elongated by this force. If no force is exerted as the wire cools, very little deformation or stretch occurs in the cool, room-temperature state and correspondingly very little contraction occurs upon heating. Up to a point, the higher the load the higher the stroke. The strength of the wire, its pulling force, and the bias force needed to stretch the wire back out are a function of the wire size of cross-sectional area and can be measured in pounds per square inch (psi). If a load of 5,000 psi is maintained during cooling, then about 3 percent memory strain will be obtained. At 10,000 psi, about 4 percent results, and with 15,000 psi and above, nearly 5 percent is obtained. There is a limit, however, to how much stress can be applied.

Far more important to stroke is how the wire is physically attached and made to operate. Dynamics in applied stress and leverage also vary how much the actuator wires move. While normal bias springs that increase their force as the Flexinol actuators contract have

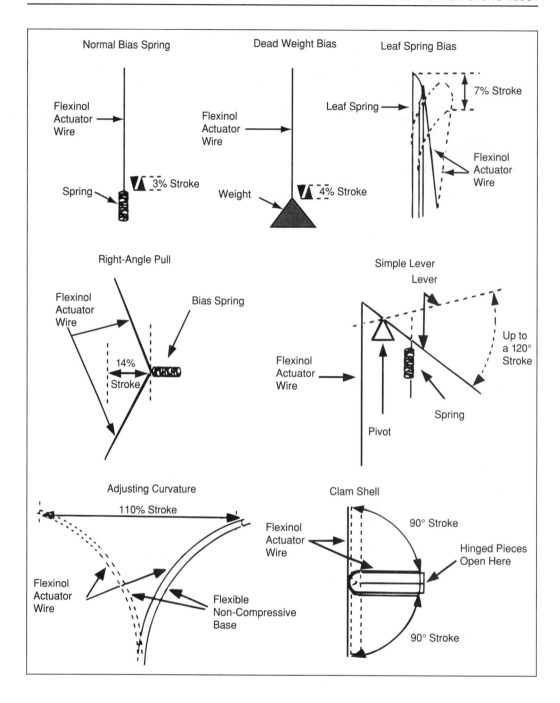

only 3 to 4 percent stroke, reverse-bias forces that decrease as the actuator wires contract can readily allow the wire to flex up to 7 percent. Mechanics of the device in which it is used can convert this small stroke into movements more than 100 percent of the wire's length and at the same time provide a reverse-bias force. The stress or force exerted by Flexinol actuator wires is sufficient to be leveraged into significant movement and still be quite strong. Some basic structures, their percent of movement, and the approximate available force they offer in different wire sizes are as follows:

Stroke and Available Force

	Approx. Stroke	.001" Wire	.002" Wire	.003" Wire	.004" Wire	.005" Wire	.006" Wire	.008" Wire	.010" Wire
Normal Bias Spring	3%	7 gm	35 gm	80 gm	150 gm	230 gm	330 gm	590 gm	930 gm
Deadweight Bias	4%	7 gm	35 gm	80 gm	150 gm	230 gm	330 gm	590 gm	930 gm
Leaf Spring Bias	7%	7 gm	35 gm	80 gm	150 gm	230 gm	330 gm	590 gm	930 gm
Right-Angle Pull	14%	2 gm	9 gm	20 gm	38 gm	56 gm	83 gm	148 gm	232 gm
Simple Lever*	30%	1 gm	5 gm	11 gm	22 gm	31 gm	47 gm	84 gm	133 gm
Adjusting Curvature	110%	.2 gm	1.2 gm	3 gm	5 gm	8 gm	12 gm	22 gm	34 gm
Clam Shell	100%	.2 gm	1.4 gm	3.2 gm	6 gm	9 gm	13 gm	24 gm	37 gm

*Assumes a lever ratio of 6:1

SECTION 2. ELECTRICAL GUIDELINES

If Flexinol actuator wire is used within the guidelines, then obtaining repeatable motion from the wire for tens of millions of cycles is reasonable. If higher stresses or strains are imposed, then the memory strain is likely to slowly decrease and good motion may be obtained for only hundreds or a few thousand cycles. The permanent deformation that occurs in the wire during cycling is heavily a function of the stress imposed and the temperature under which the actuator wire is operating. Flexinol wire has been specially processed to minimize this straining, but if the stress is too great or the temperature too high some permanent strain will occur. Since temperature is directly related to current density passing through the wire, care should be taken to heat, but not overheat, the actuator wire. The following chart gives rough guidelines as to how much current and force to expect with various wire sizes.

Diameter Size	Resistance Ohms/Inch	Maximum Pull Force	Approximate Current at Room Temperature*	Contraction Time*	Off Time 70°C Wire	Off Time 90°C Wire
.0015"	21	17 gm	30 mA	1 sec	.25 sec	.09 sec
.002"	12	35 gm	50 mA	1 sec	.3 sec	.1 sec
.003"	5	80 gm	100 mA	1 sec	.5 sec	.2 sec
.004"	3	150 gm	180 mA	1 sec	.8 sec	.4 sec
.005"	1.8	230 gm	250 mA	1 sec	1.6 sec	.9 sec
.006"	1.3	330 gm	400 mA	1 sec	2 sec	1.2 sec
.008"	.8	590 gm	610 mA	1 sec	3.5 sec	2.2 sec
.010"	.5	930 gm	1000 mA	1 sec	5.5 sec	3.5 sec
.012"	.33	1250 gm	1750 mA	1 sec	8 sec	6 sec
.015"	.2	2000 gm	2750 mA	1 sec	13 sec	10 sec

*The contraction time is directly related to current input. The figures used here are only approximate since room temperatures, air currents, and heat sinking of specific devices vary. Currents that heat the wire in 1 second can be left on without overheating it. Both heating and cooling can be greatly changed (see Section 3, Cycle Time).

SECTION 3. CYCLE TIME

The contraction of the Flexinol actuator wire is due solely to heating and the relaxation solely to cooling. Both contraction and relaxation are virtually instantaneous with the temperature change of the wire. As a result, mechanical cycle speed is dependent on and directly related to temperature changes. Applying high currents for short periods of time can quickly heat the wire. It can be heated so fast in fact that the limiting factor is not the rate at which heating can occur but rather the stress created by such rapid movement. If the wire is made to contract too fast with a load, the inertia of the load can cause overstress to the wire. To perform high-speed contractions, inertia must be held low and the current applied in short, high bursts. Naturally, current which will heat the wire from room temperature to over 100°C in 1 millisecond and will heat it much hotter if left on for any length of time.

While each device has different heat-sinking and heating requirements, a simple rule of thumb can be used to prevent overheating. Measuring the actual internal temperature of the wire across such short time periods is somewhat problematic; however, one can tell if the actuator wire is overheated simply by observing if the wire immediately begins to cool and relax when the current is shut off. If it does not promptly begin to relax and elongate under a small load when the power is cut, then the wire has been needlessly overheated and could easily be damaged. Simple visual observation is all that is needed to design measured heating circuitry.

Flexinol actuator wire has a high resistance compared to copper and other conductive materials but is still conductive enough to carry current easily. In fact, one can immerse the wire in regular tap water and enough current will readily flow through it to heat it. All of the conventional rules for electrical heating apply to the wire, except that its resistance goes down as it is heated through its transformation temperature and contracts. This is contrary to the general rule of increased resistance with increased temperature. Part of this drop in resistance is due to the shortened wire, and part is due to the fact that the wire gets thicker as it shortens, roughly maintaining its same three-dimensional volume. It makes no difference to the wire whether alternating current, direct current, or pulse-width modulated current is used.

Again, relaxation time is the same as cooling time. Cooling is greatly affected by heat sinking and design features. The simplest way to improve the speed of cooling is to use smaller diameter wire. The smaller the diameter the more surface-to-mass the wire has and the faster it can cool. Additional wire, even multiple strands in parallel, can be used to exert whatever force is needed. The next factor in improving the relaxation or cooling time is to use higher temperature wire. This wire contracts and relaxes at higher temperatures. Accordingly the temperature differential between ambient or room temperature and the wire temperature is greater and, correspondingly, the wire will drop below the transistion temperature faster in response to the faster rate of heat loss.

Other methods of improved cooling are to use forced air, heat sinks, increased stress (this raises the transistion temperature and effectively makes the alloy into a higher transition temperature wire), and liquid coolants. Combinations of these methods are also effective. Relaxation time can range from several minutes (delay switches) to fractions of milliseconds (miniature high-speed pumps) by effective and proper heat sinking. The following chart gives some idea of the effect these various methods have.

Relative Effects of Cooling Methods

	Improvement in speed*
Increasing Stress	1.2 : 1
Using Higher Temperature Wire	2 : 1
Using Solid Heat-Sink Materials	2 : 1
Forced Air	4 : 1
Heat-Conductive Grease	10 : 1
Oil Immersion	25 : 1
Water with Glycol	100 : 1

*These improvements are not cumulative on the same basis when used together.

Better cooling methods are likely to require more current or heat to move and/or hold the wire in an "on" position. In some cases one may wish to quickly turn the wire on (that is, electrically heat it until it contracts) then hold it on for some time. This will likely require a two-step driving current, with a larger current to heat the wire and a reduced current to keep it hot without overheating it. There are a number of simple circuits that will do this.

SECTION 4. MISCELLANEOUS

Cutting—Flexinol actuator wire is a very hard and anticorrosive material. It is so hard that it will damage cutters designed to cut copper and soft electical conductors. If you plan to do much work with Flexinol actuator wires, a good high-quality pair of cutters like those used to cut stainless steel wires will be a good investment.

Attaching—Attaching Flexinol actuator wires to make both a physical and an electrical connection can be done in several ways. It can be attached with screws, wedged onto a PC board, glued into a channel with conductive epoxies, and even tied with a knot. The simplest and best way is usually by crimping or splicing. With crimping machines both electrical wires and hooks or other physical attachments can be joined at once.

Flexinol wires tend to maintain the same volume, so when they contract along their length, they simultaneously grow in diameter. This means the wires expand inside the crimps and hold more firmly as the stress increases through pulling. While this works to the advantage in crimps it can be a disadvantage if glues or solder is used, as the material tends to work itself loose in those cases. Flexinol wire is a very strong material and is not damaged by the crimping process. Dynalloy, Inc. can provide wire that is already crimped at specified intervals. One can then solder or spot weld to the crimps if such manufacturing methods are preferred.

Accompanying Materials—Flexinol actuator wires work by internal resistance or other heating methods. Their temperature is often over 100°C and they often apply pressure with a high force over a small area of the device they are attached to, so it is a good idea to use temperature-resistant materials in connection with them. Such materials, if used in direct contact with the wire, will also need to be nonconductive so as not to provide an electrical path around the Flexinol actuator wire. Silicone rubber, Kapton (used to make flexible circuit boards), ceramics, and glass are good examples.

Strain Reliefs—Overstress can damage Flexinol wires by permanently stretching (or elongating) them or by reducing the stroke over which they contract. To prevent this, one should design products with strain reliefs in them. Care should also be taken to prevent manual interference with their contraction or movement as this can overstress the wire. In other words, if the device gets stuck and cannot move or is forced backward while operating, a problem can be created, breaking or adversely affecting the actuator wires' performance. Protective measures should be used against this.

Reverse Bias—Although Flexinol actuator wire moves about 4.5 percent when lifting a weight or when contracting against a constant force, one can improve this stroke by designing mechanisms that have a reverse-bias force. The bias force is the force that elongates the wire in its rubber-like martensitic phase. A reverse-bias force is one which gets weaker as the stroke gets longer. This can be done with leaf springs or with designs that give the Flexinol actuator wire a better mechanical advantage over the bias spring or force as the stroke progresses.

Performance Margin—Although very stable compared to other similar alloys, Flexinol actuator wires will permanently stretch out or strain with large cycle strokes and high stresses. At stresses below 15,000 psi, permanent strain will remain less than 0.5 percent even after hundreds of thousands of cycles. At 20,000 psi, perhaps 1 percent permanent strain will occur after 100,000 cycles, and with higher stresses proportionally more will occur.

Good engineering design dictates that one should take into account the amount of memory strain, possible small decreases in the amount of that strain during operation, and some permanent deformation of the wire during cycling if the design is to meet expectations. Pushing all performance aspects of the wire to the limit from the outset of its cycling is likely to lead to disappointment at an early stage in the product life.

Longevity Testing—Flexinol actuator wire can be overstressed and damaged even though it seems to be working. Much like actual muscles can be strained when called upon to do work above their actual capacity, the device may work in such a way that it is difficult to calculate the actual stresses involved. A good suggestion is to perform life cycle tests before assuming that a device which has worked a few times will continue to work millions more times. Fatigue that is damaging to Flexinol actuator wire will usually show up on the form of wire elongation or reduced stroke within the first few hundred strokes. As one works with the material, a feel for what is working will develop. The best rule of thumb is to use enough Flexinol actuator wire to be sure one is well within the parameters in which it can work.

Precise Positioning—Given close temperature control under a constant stress one can get precise position control. Control in microns or less is to be expected. The problem is precise temperature control. The temperature is determined by an equilibrium between the rate of heating and the rate of cooling. Heating by electricity makes control easy, but the cooling is dynamically affected by changes in room temperature, air flow, and so on. In practical terms this means precise control is usually not feasible unless one can control the heat loss or has dynamic feedback through a closed-loop system and can use this to control the heating rate.

Contact Dynalloy, Inc.—There is no practical way for the authors to include everthing that has been learned or will be learned in this short document. We have thousands of customers who call and contribute to our general understanding of typical application solutions. In most cases, we have already encountered problems that seem new to the first-time user, so whenever possible we are happy to pass on these suggestions and be of help. We want your project to succeed, so please do not hesitate to call for assistance.

SECTION 5. UNDERLYING TECHNOLOGY

Flexinol is a trade name for high-performance shape-memory alloy actuator wires. Made of nickel-titanium, these small-diameter wires have been specially processed to have large, stable amounts of memory strain for many cycles. In other words, they contract like muscles when electrically driven. This ability to flex or shorten is characteristic of certain alloys, which dynamically change their internal structure at certain temperatures. Flexinol wires contract by several percent of their length when heated and then easily elongate again by a relatively small load when the current is turned off and they are allowed to cool.

The function of the Flexinol wire is based on the shape-memory phenomenon that occurs in certain alloys in the nickel-titanium family. When both nickel and titanium atoms are present in the alloy in almost exactly a 50/50 ratio, the material forms a crystal structure that is capable of undergoing a change from one crystal form to another (a martensitic transformation) at a temperature determined by the exact composition of the alloy. In the crystal form that exists above the transformation temperature (the austenite) the material is high strength and not easily deformed. It behaves mechanically much like stainless steel. Below the transformation temperature, however, when the other crystal form (the martensite) exists, the alloy can be deformed several percent by an uncommon deformation mechanism, which can be reversed when the material is heated and transforms. The low-temperature crystal form of the alloy will undergo the reversible deformation fairly easily, so the memory strain can be put into the material at rather low stress levels.

The resultant effect of the shape-memory transformation of the Flexinol wire is that the wire can be stretched about 4 to 5 percent of its length below its transformation temperature by a force of only 10,000 psi or less. When heated through the transformation temperature, the wire will shorten by same 4 to 5 percent that it was stretched, and can exert stresses of at least 25,000 psi when it does so. The transformation temperature of the NiTi alloys can be adjusted from over 100°C down to cryogenic temperatures, but the temperature for the Flexinol actuator wire has been chosen to be 60 to 110°C. This allows easy heating with modest electrical currents applied directly through the wire, and quick cooling to below the transformation temperature as soon as the current is stopped. Heating with electrical current is not required, but it is perhaps the most convenient and frequently used form of heat.

Flexinol actuator wires' prime function is to contract in length and create force or motion when heated. Ther are limits, of course, to how much force or contraction can be obtained. The shape-memory transformation has a natural limit in the NiTi system of about 8 percent. That is the amount of strain that can occur in the low-temperature phase by the reversible martensitic twinning that yields the memory effect. Deformation beyond this level causes dislocation movement throughout the structure, and then that deformation is not only irreversible but degrades the memory recovery as well. For materials expected to repeat the memory strain for many cycles, it is best to use a cyclic memory strain of no more than 4 to 5 percent, and that is what is recommended with Flexinol actuator wire.

The force that Flexinol actuator wire can exert when heated is limited by the strength of the high-temperature austenitic phase. The phase transformation, or crystal change, that causes the memory effect has more driving force than the strength of the parent material, so one must use care not to exceed that yield strength. The yield strength of Flexinol's high-temperature phase is more than 50,000 psi, and on a single pull the wire can exert this force. To have repeat cycling, however, one should use no more than two-thirds of this level, and forces of 20,000 psi or less give the best repeat cycling with minimal permanent deformation of the wire.

Index